Electronic Imaging Systems

Design, Applications, and Management

Electronic Imaging Systems

Design, Applications, and Management

Don Avedon
Joseph R. Levy

McGraw-Hill, Inc.
New York San Francisco Washington, D.C. Auckland Bogotá
Caracas Lisbon London Madrid Mexico City Milan
Montreal New Delhi San Juan Singapore
Sydney Tokyo Toronto

Library of Congress Cataloging-in-Publication Data

Avedon, Don M. (Donald M.)
Electronic imaging systems: Design, applications, and management / by Donald M. Avedon and Joseph R. Levy.
p. cm.
Includes index.
ISBN 0-07-002484-7
1. Imaging systems. I. Levy, Joseph, 1934- . II. Title.
TK8315.A94 1994
651.5'0285'66—dc20 93-20955
CIP

1 2 3 4 5 6 7 8 9 0 HAL/HAL 9 9 8 7 6 5 4

ISBN 0-07-002484-7

The sponsoring editor for this book was Dan Gonneau, the editor was Marianne Krcma, the designer was Jaclyn J. Boone, and the production supervisor was Katherine G. Brown. This book was set in Century Schoolbook. It was composed in Blue Ridge Summit, Pa.

Printed and bound by R.R. Donnelly of Harrisonburg.

Joseph Levy's portion of this book is dedicated to the memory of his high school classmate, Dr. Norma Lemon.

Contents

Acknowledgments

Joseph Levy would like to thank Hewlett-Packard and Pentax Technologies for providing scanners for this project.

Preface

Electronic imaging is one of the hottest topics in computer circles, but just try finding a book about it! The information tends to be buried inside limited-circulation reports and other obscure resources. I could not, for example, find a single book on the subject at the downtown branch of the Public Library of Cincinnati and Hamilton County, which is considered one of the best public libraries in the county. Nor could I find any books in my local CompUSA or MicroCenter, two superstores that pride themselves on their professional collections. (The closest I came were some how-to books tied to specific products, such as individual scanner models.)

When you finally find a book, it is usually a technical tome written for experts in the field. This book, on the other hand, is a professional primer for managers and computer users facing the electronic imaging revolution for the first time (or who are thinking about getting a system). It was created for the vast majority of professional imaging users, who are, as far as imaging is concerned, non-experts. In other words, this book is intended as a first step into the imaging world—the logical extension of Don Avedon's introductory electronic imaging lectures combined with a novice's viewpoint. This book is designed to leave you with a working knowledge of electronic imaging, a sense of how it fits in with other trends in the computer industry, and hard-to-find contact information for imaging resources.

Once you find out how much time and effort an imaging system can save you, you might want to learn even more about the area. If so, contact the Association for Information and Image Management (AIIM) in Silver Spring, Maryland. AIIM is the nexus connecting all facets of the electronic imaging industry. Also, review and possibly subscribe to Datapro's Document Imaging Systems series. Datapro, in Delran, New Jersey, can provide you with up-to-date, full-length reports on all of the intricacies of the electronic imaging system, including in-depth analyses of vendor offerings and evaluations of upcoming trends mentioned in this book.

The electronic imaging field is changing constantly. With the help of this book, you can follow those changes.

—Joe Levy

Chapter

1

Electronic Imaging: Understanding the Basics

"It is said that humans will build a fire for one of two reasons: either because they are cold or because they have a match. Today the information technology community is ablaze with imaging because someone struck the match of worldwide competition. This competition demanded not just incremental process improvements but a wholesale rethinking of the way business is conducted. What emerged was a confluence of technologies that enabled electronic image management (EIM) to flare up from its well-established base in micrographics to become the most dynamic element in a 3.5 billion dollar industry."

AIIM White Paper

What exactly is *electronic imaging*? The term is new to most of us. The simplest way to think about electronic imaging is from the viewpoint of the user. Odds are, after all, that you're not an electronic imaging professional, but rather a user of the increasingly widespread electronic imaging technologies. From a user's viewpoint, the easiest way to understand electronic imaging is as the representation of a paper document on a computer screen.

Actually, electronic imaging involves a systematic approach to presenting information that way. An imaging system captures and stores records in *digital* (electronic) form. It also retrieves and disseminates them. The simplest electronic imaging system has four hardware components:

- An optical scanner for creating a digitized image of a document
- Optical disks for storing the images
- A disk drive for recording the images and retrieving them
- A high-resolution terminal for viewing retrieved images

Many people who think of imaging concentrate on the scanning phase. Indeed, that part of the process is the most dramatic. Perhaps you have seen a demonstration of a scanner copying fine art or other graphic materials. The latter phases are also important, however, because no one wants to lose their documents once they have been scanned and stored. One of the reasons for having electronic imaging in the first place is that conventional storage techniques lose information easily. Indeed, Wang Corporation found that the average executive spends three hours each week looking for lost information, while his or her assistant spends another four to five hours looking for the same information. Most of us know what it's like to go through the time-consuming ritual of "Now where did I put . . .?" Indeed, most "paper-intensive" business operations end up misfiling three out of 100 records.

After an image is scanned, it is classified or *indexed*. Usually, this information is entered by keyboarding. However, bar coding and *OCR (optical character recognition)* technology can reduce manual input. Image indexing can range from serial numbering to long, structured descriptions. The information is stored in databases specifically designed for image data. Usually, images are compressed into standard file formats, such as TIFF (Tagged Image Field Format) or CCITT Group 3 or Group 4, before they are entered.

Indexing is a key part of almost any electronic imaging system. Each document's index is the handle for locating it inside an imaging system. Indexing errors in an electronic imaging system cause many problems. Looking for a lost piece of paper in old-fashioned filing cabinets is easier and poses fewer security problems than searching for a document in a database. Without proper indexing, moreover, electronic documents can be lost forever.

The choice of image storage medium is influenced by the length of time the images need to be stored. Magnetic storage is best when there is a low volume of material that requires a high retrieval rate. Optical storage works well for high-volume, high-retrieval rate situations. In general, microfilm and optical storage are preferred for documents that need to be stored for a long time. Users retrieve images by searching on keywords.

Image decompression usually takes place at the workstation screen where the images are displayed. Sometimes just looking at a document on screen isn't enough, however; you might want to see a printout of a document. For example, you might need a copy of your birth certificate to get a driver's license. Your city or county has its birth certificates filed on computer, but that wouldn't do the state-run Department of Motor Vehicles any good. That is why many electronic imaging systems have printers. You can go to the city department and ask them to print out a copy of your birth certificate and present it to the state. In many cases, the printout produced by a printer in an electronic imaging system will look just like an original document before it has been scanned.

File-and-Find Systems

A basic electronic imaging system—scanner, disk drive, and terminal—is also known as a *file-and-find system*. The easiest way to understand this system is to think of it as a substitute for a filing cabinet. A filing cabinet stores documents in files. An electronic imaging system stores electronic documents. Either way you start out with information such as words, tables, pictures, and store them for later retrieval. Filing cabinets store original documents. Electronic imaging systems store computerized copies of those documents, which are generally scanned in. An electronic imaging system also has the advantage of being able to capture three-dimensional objects. A special piece of equipment known as a *frame grabber*, for example, digitizes images captured by a video camera.

Electronic images can be stored in one of the three ways:

- Black and white images
- Grayscale images
- Color images

Black and white images are the simplest and require the least amount of storage space. Line drawings, typed text, and handwritten messages are examples of information stored in black and white. They are stored as *binary codes*; the scanner sees the millions of points that make up the image on the document as either black or white. Literally, there are no grays. This information then is stored inside a computer bit. This bit is like a switch that can either be turned on to show white or turned off to show black. To be more precise, the bit can count, but only to one, starting from zero. It only understands two numbers, *1* and *0*. The computer assigns the value *1* to any tiny piece of information it sees as black and *0* to any tiny piece it sees as white.

Continuous-tone images, such as photographs, are more complex. The computer turns the scanned photograph into a series of black and white dots. Each dot appears on the screen as a pixel, also known as a picture element or *pel*. The more bits per pixel, the more tones that can be produced. In other words, the more blacks and whites you have in an area, the greater the number of types of grays you can display. A four-bit pixel will turn out 16 shades of gray. On the other hand, 256 bits of gray can be produced by an eight-bit pixel. You need 24 bits per pixel to display color.

Color reproduction hardware is still in the development phase. Analysts hope that by the end of the decade, the industry will standardize color technologies and have color calibration as part of a dedicated operating system function. Today, however, a color photograph will not look the same when you input it in a color scanner and when a printer prints it out. Color images require triple the disk storage requirements of monochrome images and are expensive to reproduce.

Text management, however, extends way beyond file-and-find capacities. *Hypertext systems* allow users to jump directly to cross-references by linking conceptually related textual material. *Electronic publishing* systems consist of products assisting the visual and typographic layouts of specific documents. *Document assembly systems* recombine existing text blocks into new "virtual" documents that can be published on demand. *Expert systems* consist of products capable of making informed decisions based on accurate internal representation of knowledge as well as executable rules for applying that knowledge. (Expert systems also cover transaction processing and technical documents.)

The relations of expert systems and electronic imaging has been of particular interest to analysts who foresee a marriage of imaging technology and expert systems that increases productivity. As expert systems become more integrated with imaging systems, imaging systems, in turn, are becoming more common. Indeed, expert systems can be integrated with imaging to varying degrees and at multiple levels in the same way that imaging applications can be loosely or tightly integrated with existing hardware and software architectures. The result is a continuum, with imaging and expert-system technologies on the low end (using expert systems to select data from scanned images), cross-reference images based on content and probable use in the middle, and balanced workflow in operational imaging systems at the high end.

With a low focus on sequential processing, first one technology performs a function, then the other, such as data-preparation functions on the front end and selective-reporting functions on the back end. High level integration, such as dynamic-data translation (the movement of data between applications), involves significant interaction between the technologies. Regardless of the level, the goal is seamless integration.

Almost every major industry can provide examples. The largest market category of integrated imaging, however, is file-folder applications. These applications, typically accounting and financial functions, are common because they can be applied independently in all vertical industries. Integrated imaging and expert systems are becoming increasingly common in government and legal applications because of the vast volume of paper there.

New Technologies Converge

Of course, electronic imaging is not the first technology to address the information explosion. Micrographics technology was there first. Actually, electronic imaging grew out of the marriage of computers and micrographics. Manufacturers, government agencies, health-care providers and financial service organizations have been using micrographics as a way of storing information for decades. Indeed, the Association for Information and Image Management (AIIM), the trade organization for the electronic imaging business, owes its existence to the National Microfilm Association founded in 1945.

Microfilm

Of all the types of micrographics, most people are probably the most familiar with microfilm. In conventional microfilming, the reduced image of a document is recorded photographically onto high-resolution film. The key advantage of microfilming is that it can provide an opportunity to automate the entire recording and filing process by allowing the storage, retrieval, and use of images within the same physical location. Many firms use microfilm as an efficient way of storing records, saving both space and time.

However, conventional microfilming also has its drawbacks. Some of the key limitations of conventional microfilming include a lack of immediacy and an inability to update files. Because of its lack of immediacy, you cannot judge the quality of microfilm images right away. Updating problems are caused by the inability to record over the same microfilm. However, the use of jackets (cut strips of roll film arranged in parallel transparent sleeves) and updatable microforms can sometimes resolve the problem of changes and additions.

Reel-based microfilm and microfiche represent only one category of micrographic products called *standalone micrographics*. The other two categories are *computer output to microfilm* (*COM*) and *computer-assisted retrieval* (*CAR*). Computers have been sending output to microfilm and microfiche almost as long as they have been printing out to paper. Computers have also been used to assist in the retrieval of micrographic images for years.

As with any technological solution, electronic microfilming systems have both champions and critics. Critics complain about the start-up costs, which are high and make any design errors expensive to correct. Advocates point out that electronic microfilming offers several advantages over conventional microfilm systems, including the following:

- Images can be displayed after capture, providing immediate availability of the captured documents, as well as real-time quality control.
- Captured images can be sent to several users at one time.
- Because retrieval of computerized records can be prescribed independently of the recording sequence, adding a record to a file is simple.

Ultimately, electronic imaging and microfilm are not adversaries. Interest in conventional microfilm is continuing in the electronic age. One of the strongest reasons for a continued interest in microfilm and microfiche is economics. Conventional microfilm is inexpensive and standardized. It is also acceptable for legal purposes and easy to read. Computer-aided retrieval has made finding images simpler.

The growth of electronic imaging results directly from the development of several other diverse technologies. Electronic imaging, for example, is possible only because of the development of high-density mass-storage media. A full-page image requires at least 50,000 bytes (with eight bits per byte) and perhaps as many as one million bytes. Word-processed text is less space-consuming, requiring 4,000 to 5,000 bytes to store a full page of text. However, storing a large number of pages still poses a space problem. Fortunately, mathematical formulas known as compression algorithms have been developed over the years to get the most from existing storage space.

An optical disk with its economical, multigigabyte capacity is the most essential element of most EIM systems. They are used in conjunction with magnetic disks. These disks are described in detail in chapter 3.

Microprocessors and OCR

Another important advance has been the development of microchips. Most people with a technical computer background already know quite a lot about microchips. While no one really needs to know how a microprocessor works to run a computer, any more than someone needs to know how a piston works in order to drive a car, the best way to understand a microprocessor is to think of it as a small computer. Intel Corporation developed the microprocessor, or computer on a chip, which has led to the creation of so-called fourth-generation computers.

A microprocessor stores information in different ways, depending on how it is to be handled. Information that you, the user, can alter is stored in random-access memory, *RAM*. On the other hand, the actual operating program is normally stored in read-only memory, *ROM* (or in a permanent logic array). Information that is stored in RAM is lost when the power supply is disconnected. Information stored in ROM, on the other hand, is retained even if the power is interrupted. RAM is short-term memory, ROM long-term memory.

Datapro, the marketing research firm, recently studied the microprocessor choices of 100 leading imaging vendors who used PCs. They found both IBM PC/AT and PS/2 family as well as more specialized IBM RS/6000 and Sun Microsystem's RISC processors being used by imaging systems. Any improvements in chip technology, such as increased speed, will filter down to imaging systems.

Optical character recognition (OCR) technology has greatly increased the functional ability of imaging systems. OCR enables computer equipment to "read" and input words. First introduced some 30 years ago, the earliest forms of OCR required special inks, typefaces, and papers. Despite the inconvenience, many banks adopted OCR systems to handle check processing limited to a few typewriter fonts. In the 1970s, $50,000 standalone systems reached the market. They were, however, the domain of high-volume operations and service bureaus and still had very limited capacities. Today OCR systems have advanced to read many typefaces and font sizes. Some systems even have some capacity to read handwritten letters.

These more advanced systems that can learn new type styles are known as *intelligent character recognition* systems. Some ICR systems can find specific data on a page, such as new input, and ignore others, such as background information. The development of OCR technology as a whole and ICR technology in particular, however, has been slowed by accuracy problems. Even today OCR engineers continue to work on ways to increase accuracy.

Meanwhile, another important development in the industry has been the growth of workstation technology. A *workstation*, of course, is a high-powered microcomputer system, including a high-resolution monitor, substantial memory, and specialized hardware, such as a decompression board. A workstation is really a communications center that offers functions such as word processing, electronic mail, access to mainframe computers, access to online retrieval services, electronic spreadsheets, and administrative tools, including calendars, reminder files, automated telephone directories and other functions. In 1981, workstation manufacturers shipped $1.6 billion worth of workstations. By 1985, that climbed to $3.1 billion.

One of the biggest areas of improvements in workstation design over the years has been monitors and their compatible cards. Monochrome display adapters (MDAs) allowed for text displays but no graphics. Of course, they were black-and-white. The Hercules graphics card (HGC) was black-and-white, but could display graphics. Most of today's monitors have color capacity as well. The color graphics adapter (CGA) was the first color standard to gain widespread acceptance, displaying up to 16 colors at once in low resolution. The enhanced graphics adapter (EGA) raised the ante to 64 colors. The video graphics array (VGA) supports 256 colors. Super VGA (SVGA) supports 256 colors or more in high resolution.

While there are no formal standards for electronic imaging monitors, many analysts will tell you that a resolution of 1024×768 pixels is an acceptable standard for document imaging displays. Some manufacturers market displays of 120 dpi or higher. You can also get video accelerators to increase resolution and speed up color image processing. Screen sizes necessary for electronic imaging generally range from 17 to 21 inches. You need a 19-inch display to present a two-page document. Twenty-one inch and larger screens usually are used for CAD/CAM systems and medical imaging or other demanding applications.

Software has also advanced rapidly in recent years. As chapter 5 details, general changes in the software industry filter down to imaging applications. The most popular operating system in use on PCs today is DOS. DOS allows you to boot up your computer and run applications programs. One popular competitor of DOS is OS/2. OS/2 allows for multitasking, which means that one computer can do more than one job at a time. Windows is another popular program because it allows PC users to work with a mouse and a screen that resembles a Macintosh configuration. Windows NT is the latest version of Windows and includes many of the features found in OS/2.

Applications programs today that run under DOS, OS/2, and Windows include the following:

- Word processing for use in preparing documents
- Spreadsheets for accounting work
- Graphics for producing art
- Databases for storing large quantities of information

Imaging operating systems need not be limited to DOS and OS/2. Windows NT, many analysts feel, has a strong future as an imaging system. Windows NT runs on Intel microprocessors, Digital Equipment's Alpha series and MIP's Computer System RISC processors. UNIX, a mainframe operating system that has migrated to PCs, is also a part of the electronic imaging world. Sunsoft's Solaris 2.0 operating system is also important to imaging. It will run on SPARC microprocessors and PCs.

Electronic imaging software is constantly evolving. Today's EIM packages include software for

- Indexing
- Scanning
- Image enhancement
- Image file conversion
- Image visualization
- Editing and manipulation
- Video image capture and processing
- Image compression and decompression
- Printing
- Imaging development toolkits
- Libraries

You can find a variety of off-the-shelf systems designed to manage document images. Many of them are general-purpose systems, while others facilitate customization. Many also allow users to manage other document types, including word processing documents and data downloaded from mainframes. The programs usually come with their own database, image-indexing method, image utilities, image device support, and application interface to I/O services. Popular systems include IBM's ImagePlus, FileNet's Workflow Business System and MicroDynamics' MD MARS.

Frame grabbers

Another important hardware development in the evolution of electronic imaging technology has been the creation of frame grabbers. These add-in boards for PCs "grab" single frames from video presentations for use in online or printed documents. A frame grabber occupies one or, in some cases, two slots in a personal computer and includes input jacks to connect it to a video source. In most cases, the video source is a VCR or videocamera. To acquire an image for use on the computer, the frame grabber takes a video signal, then uses digital-to-analog converters and circuitry to align the video signals with the timing of the computer display so that associated software can capture, compress, and save the frozen video image. Frame grabbers are used for medical imaging, military intelligence, scientific research applications, and desktop publishing applications such as training manuals, newsletters, and reports.

Scanners

The simple reality is that imaging could not exist without some form of optical scanners. Scanner sales are climbing. The marketing firm BIS Strategic Decisions forecasts a 31 percent annual growth rate for scanners through 1995. BIS expects one million color scanners—half of all scanners shipped—to be color by 1996. Market growth will take place, analysts say, with widespread acceptance of the ease of scanner use.

A few years ago Hewlett-Packard was the only large company manufacturing scanners. Today, however, Xerox, Ricoh, Bell & Howell, and Fujitsu are very active in the scanner market. One of the leading manufacturers of scanner hardware and software is Calera. Calera has designed its WordScan software and other products to take advantage of many platforms, such as the Macintosh environment and SPARC workstations.

Although scanners vary widely in price, size and models, they all have the same basic components. The scanner's sampling device, usually a charge-coupled device (CCD), passes information onto a controller. (By the way, CCDs were originally used to generate images beamed down to earth from satellites.) The CCD usually consists of a semiconductor chip with an array of light-sensing elements. A lamp illuminates the page during scanning. A lens reflects and concentrates the light onto the chip. As the feeder system moves, changes occur in the pattern of reflected light. The CCD returns a high voltage corresponding to white areas, low corresponding to black, and gray in between. Other scanners use light-emitting diodes (LEDs) or lasers.

The controller, which is located either in the host computer or scanner, interprets electronic signals from the sampling device and controls mechanical movements. The scanner also uses software to perform its functions, either loaded separately or as part of the operating system.

Ultimately, the scanner market can be divided into three segments: low- to mid-volume scanners, mid- to high-level scanners, and very low-volume desktop scanners. Low- to mid-volume scanners are used where less than 3,000 pages are scanned in one location. The current base price for the mid-volume Bell & Howell Copiscan II Model 2135 is $7,395. At the higher end of the scale, TDC offers a DS-2610W with an autofeed for $30,500. You can get a desktop unit such as the ScanJet IIc for less than $1,600.

Fax machines

The fax or facsimile machine is not really a new invention at all; it is an old technology reborn with a digital slant. Today's digital machines have higher-resolution scanning, faster transmission, lower prices, and cleaner, more detailed outputs than their predecessors. However, analog faxes have been used in newspaper publishing for almost 30 years.

Fax machines are important in the imaging fields because faxes can sometimes take the place of scanners. Because the compression standards used by facsimile are the same as those used in image processing, the two technologies are compatible. Fax machines work by using a two-phase process. A sending machine will:

1. Scan an original document line-by-line.
2. Compress the scanned data.
3. Transmit the data via telephone lines.

A receiving fax machine will:

1. Decompress the scanned data.
2. Reproduce the document line-by-line using a thermographic or xerographic printing process on bond paper.

While this process tends to be slower than scanning with a scanner, fax capabilities are an important part of electronic imaging. A 1993 Datapro survey found that four out of five imaging systems could send and receive facsimiles.

Many PCs come equipped with fax boards. These boards act like a fax machine without an input scanner or hardcopy output device. They transfer files directly into a computer, a feature that is often useful in imaging applications. Some fax boards enable the assembly to fax transmissions from separate documents or graphic images. These units can track multiple fax transmissions or receipts.

Recent years have seen the development of multifunction or hybrid products. For example, Umax's ReadStation incorporates a scanner, automatic document feeder, dedicated computer, OCR and ICR software. These types of products will become more available as their sophisticated software becomes more refined. Eventually, machines that incorporate scanning, facsimile, and printing technologies could be considered the best answer for small business office needs.

The Bandwagon Begins

The need for electronic imaging has arisen as a solution to pressing business and organizational problems. One of the key reasons people turn to electronic imaging is that it is efficient. Time is saved by the elimination of paper shuffling. Information industry estimates show that a typical worker performing in a typical department, such as accounts receivable, will take 12 minutes to process a document. Nine of these 12 minutes are spent searching for, retrieving, and refiling the document. The other three are spent processing it. Electronic imaging management dramatically decreases the time required for nonproductive operations—by as much as 75 percent.

Christopher Locke, author of the Datapro report "Foundations for Document and Information Management," points out that organizations are drowning in information: "Regardless of whether needed textual information is on disk or paper, today it is clearly beyond the ability of most organizations to adequately capture, organize, and evaluate the contents of the knowledge embedded within such texts and to shape informed decisions based on such an analysis. Many companies can no longer afford to depend exclusively on primarily programmer-oriented database technology to gain control over these crucial sources of management information." Locke contends that organizations today have a choice of updating their technologies or facing the "potentially disastrous loss of critical organizational knowledge."

In the end, the primary reason for the growth of all EIM systems may derive from the burgeoning economic impact of information technology. Today the information industry dominates our economy. Companies involved in producing the technology, equipment, systems, and supplies for the creation, use, storage and retrieval of information account for almost 50 percent of the gross domestic product of the United States.

"It is now common that approximately nine of every ten dollars spent to process information in today's economy is spent to pay the salaries of the people who do the work," one analyst maintains. "Never before has it been so important to search for ways to make it possible for individuals who process information to work smarter instead of harder." Paperwork costs $100 billion in the U.S. annually, in addition to the $166 billion cost of filling out government forms.

The sheer volume of paper documents is another driving force behind the creation of electronic imaging systems. AIIM estimates that American businesses generate more than 2.7 billion sheets of paper that go into file folders every day. Indeed, a typical five-drawer filing cabinet with letter-size file drawers, for example, will hold 22,500 documents. A total of 1.6 trillion pieces of paper circulates through the nation's offices each year. This adds up to 22.8 million tons of office paper annually. The number of active files, meanwhile, is growing 25 percent annually, so that the number of corporate files double every 3.5 years. In addition, record costs are doubling every year.

Archiving and storage was one of the most prominent early applications of electronic imaging. By 1989, the Association for Information and Image Management reported that this segment accounted for 39 percent of all electronic imaging management product revenues (approximately $1.6 billion). Today that segment has fallen to 3 percent, however. Other important applications, according to a 1992 AIIM study, are as follows:

- Transaction processing—50 percent
- Indirect—25 percent
- Historical (active) records processing—13 percent
- Technical drawings management—6 percent
- Electronic publishing—1 percent
- Other—2 percent

Early systems emphasized the electronic replacement of paper storage and retrieval systems. The users thought of them as little more than electronic filing cabinets. This primitive thinking was matched by a technology sometimes described as "fragmented." Customers have tended to look for solutions to specific project problems, rather than thinking in terms of creating reusable systems. The result has been that almost half of all electronic imaging budgets go for support rather than one-size-fits-all hardware and software.

Service bureaus, which provide equipment and technology on a time-lease program, have led the way towards adopting electronic imaging techniques. Even now, much of electronic imaging is in the hands of service bureaus and independent consultants. For example, Efficiency, Inc., a firm identifying office automation needs and finding cost-effective solutions, has concentrated heavily on scanning. Their clients for the Pentax IQ Scan have included

- Boeing Aerospace/Command Media Division, where they scanned hard copies of old manuals into Microsoft Word and uploaded the data into a minicomputer
- Swedish Hospital, where radiology manuals were scanned into an online database
- Group Health Cooperative, where lengthy reports received from other agencies were converted to a format usable with PCs without rekeying
- An independent insurance adjuster, where multipage hardcopy client inventories were captured for use by this individual's PC

The climb towards widespread acceptance of electronic imaging, however, has been gradual. Even these days, many people seem to be standing at the electronic imaging crossroads, asking, "Should I do it now or wait?" Executives, versed in the art of risk aversion since business school, often take a wait-and-see attitude. In the past, buyers have been cautious for a variety of reasons. Some wanted to see prices go down. Others were worried about smaller vendors going out of business. On the other hand, vendors have been pushing rapid market growth, if only to recoup research and development costs.

Cost consciousness has become a fact of life for all MIS systems working in the 1990s. In part, this new cost consciousness is fallout from widespread errors in the computer industry. To put it simply, some people overbought in the past. Results did not justify the cost incurred. If anything, there was too much enthusiasm for the idea of office automation, the process that allows organizations to handle information flow in a much faster way than they could by hand.

In practice, however, results are often not impressive. Office automation is not always a panacea for promoting efficiency. Operational bottlenecks can arise. Occasionally new systems are operationally cumbersome, cannot respond to changes in the business environment, or become quickly obsolete.

In many instances, anticipated savings do not materialize or are offset by new costs. Unfortunately, systems promoters have sometimes raised false hopes. Early concepts, such as electronic data processing, and later ones, like management information systems, promised that the use of computer-based systems would reduce overall costs by improving the transfer of information. While many of these promises have been left unfulfilled, most of the time the flaw has been in the area of poor design. Such mistakes frequently reflect an inability of system users and designers to articulate their needs, requirements, and constraints to each other. Designers end up mechanizing preexisting conditions rather than providing reasonable solutions.

A poorly designed computer system can actually decrease overall productivity, even while it increases the efficiency of part of a system. Indeed, improvements in isolated routine activities do not always result in overall operational gains. In some cases, gains in one task might mean lower performance in other areas. Under these conditions, net productivity remains low. All too often low performance is only the last in a series of frustrations that include cost overruns and delays in implementation.

Training is one of the keys to making electronic imaging and other office automation efficient. This opinion was strongly affirmed in the classic (albeit controversial) "Study On Managerial and Professional Productivity" conducted in the late 1970s by the consulting firm of Booz-Allen & Hamilton. That survey recorded between 3,000 and 8,000 time samples in each of 15 major industries and studied the results of more than 100 person-months of analyses and evaluation. The conclusion was that to better meet their automation needs, especially those essential to profit improvement, organizations needed to support their "knowledge workers," who spent an average of almost 50 percent of their time in meetings and telephone calls. Strong capital investment is also paramount to making automation systems work, Booz-Allen & Hamilton concluded.

Against this backdrop, the bad news for EIM was the 1990s began with a round of penny-pitching budget cuts despite increased user demands. The marketing research firm Index Group predicted that MIS departments lucky enough to get increases would see hikes averaging only 5.3 percent instead of the 10 and 15 percent increases common in earlier years. Banking, airlines, and manufacturing systems all faced cuts of at least 10 percent.

As a result, MIS systems chiefs have been scrambling for more creative ways to stretch their dollars. Among the strategies used are these:

- Trimming staffs
- Reorganizing departments
- Off-loading tasks to other work groups
- Contract services and outsourcing
- Attacking software maintenance
- Seeking competitive bidding

Industry observers watched in awe as these new policies broke years of managerial traditions. Large companies, usually the best targets for high-end software, have started experimenting with inexpensive shareware for the first time. Both the *Los Angeles Times* and Caterpillar Inc.'s Computer Technology Center have decided to standardize word processing with PC-Write, a package from Quicksoft, costing $14 per user. Other organizations have manipulated their contracts. Kindle Healthcare Products, for example, claims to save $120,000 per year by replacing an outside electronic mail contract with an in-house system.

Buying and selling used equipment has also become a popular business option. The National Computer Exchange, a brokerage for used computer equipment, found that its business rose 250 percent in 1990 alone. Meanwhile, large organizations are upgrading existing PCs, putting in new CPUs and replacing boards rather than purchasing new units.

Meanwhile, a new business philosophy is starting to emerge, contending that traditional cost analysis greatly underestimates the value of electronic imaging. As Debra Sheer Haverson explained in her article "The Imaging Makeover," *Midrange Systems* (October 29, 1991): "During the past few years, imaging has undergone a dramatic change in terms of user perception. When first introduced, users viewed imaging—the use of computer systems for working with images as well as data—as an expensive, glossy technology that only large, wealthy companies can afford to use. Today companies are considering it a necessity for surviving in a competitive business environment."

The cost of electronic document imaging still is undeniably high when users look at the prices of required hardware and software. However, many companies have found that strategic business and productivity benefits enable them to justify million-dollar expenditures. These benefits include a reduction of tedious, error-prone data entry tasks, improved response to customer and supplier inquiries, simultaneous access to documents by multiple users, and the ability to reduce the amount of space required for paper storage.

Nowhere is image cost-consciousness greater than in the banking industry. Mainframe-based systems can cost between $2 million and $50 million. Some banks, such as the Bank of New England, have dropped out of the development process, citing the tremendous price tag. However, bankers are finding ways to manipulate check processing technology to deliver at least some of the advantages of imaging in a cost-effective way. Central Fidelity Bank of Richmond, Virginia, and First Interstate Bank of Washington, Seattle, now offer imaged check imaging.

How can these cost-conscious banks afford imaging? The solution lies in using highly efficient software and lower-cost microcomputers. The software necessary to produce consumer statements on PCs costs far less than complete back-office systems for item processing. The Bank of Butterfield International in the Cayman Islands, for example, uses Image Networking Corporation's INC software to produce imaged check statements much like the ones at Central Fidelity Bank. The PC-based system can cost as little as $38,000 and can handle up to 50,000 checks per day. The Bank of New York, on the other hand, uses the same system to print the images of checks from its wholesale lockbox operations. Photocopying these checks cost $.0007 per item; imaging can reduce that cost by up to $.0005.

Additional software can also increase productivity (and provide cost savings). For example, Manufacturers Hanover has more than doubled its imaging capacity from 700 to 1,800 documents an hour thanks to a software system that not only scans but also creates an image database and a transaction database that remain related throughout processing. Processing is speeded at the key entry phrase because operators can key in dollar amounts directly as they open onto the screen.

Major vendors

Prices, meanwhile, have come down. Many vendors have gone out of business, but many more have not. The leaders in the computer industry have entered the imaging field, starting in the middle of 1980s. Some key players in the electronic imaging field include:

- Digital Equipment Corporation
- Wang
- Hewlett-Packard
- IBM
- Eastman Kodak
- NCR
- Unisys
- Xerox
- FileNet

Although a relatively minor player in the office segment of document imaging, Digital Equipment Corporation (DEC) has developed both hardware and software aspects of EIM. Their Compound Document Architecture (CDA) was developed to support images as one of its data types. This architecture was designed to support the seamless transmission and integration of several data types. Further, DEC strategy emphasizes offering image-capable applications, providing tools to third-party developers to create application-specific image-capable hardware.

In May 1991, DEC introduced DECimage EXpress for VMS. This easy-to-use document imaging system allows an organization to electronically capture, index, store, display, print, and distribute document images. The product is targeted to businesses that handle paper forms and invoices. The technology is based on Digital's VAX VMS platform and complies with Digital's Network Application Support (NAS) software for applications integration in multivendor environments. DECimage EXpress is ready to use without any special programming and uses application program interfaces (APIs) for integrating management functions, such as image capture, indexing, and storage into new or existing applications. Entry-level systems can cost under $10,000. For more advanced systems, depending on the configuration used, the cost per seat of a system can be as low as $20,000. Release 2.0 and 2.1 have added support for PC workstations and fax.

DECimage EXpress is sold through OEMs, VARs, and a direct sales force located in 200 regional sales offices. It is sold as three separate components:

- The DECimage EXpress Application Development Environment (DADE)
- The client component, which provides full imaging program capabilities in the standalone mode

Xerox Imaging Systems (XIS) develops document input products that use its intelligent character recognition software. These applications are used in electronic publishing, commercial typesetting, and technical documentation. XIS products runs on UNIX workstations, IBM PCs, and Macintosh systems.

Wang launched the Wang Integrated Image System (WIIS) in 1989. They also created an Open-Image Windows software development kit for the development of WIIS applications for personal computers using Microsoft Windows. By the end of 1989 Wang spokespeople reported they had almost 400 customers, including the pioneering American Express account. Wang has since deemphasized WIIS and has discontinued the manufacture of the hardware required for WIIS.

Meanwhile, Wang continues to emphasize imaging as it climbs out of Chapter 11. Its Open/image platform incorporates RS/6000, IBM PCs, Wang RISC Series, and Wang WIIS VS as imaging servers. Although the products have earned praise, no one is sure whether they will turn Wang around.

Hewlett-Packard is no longer a major player in the imaging market. The company once targeted banking, insurance, pharmaceutical, and health-care markets as well as government departments. The company is known for both high-quality products and strong support. In addition to a significant installed base in business, Hewlett-Packard is well-known for its engineering products.

The product line for imaging was known as HP's Advanced Image Management System (AIMS), based on the company's own Unix client servers and PC microcomputers. These products allowed workflow and case management from software from Recognition Equipment and Plexus software to be used with the AIMS family. Hewlett-Packard, however no longer markets the AIMS system.

Plexus Software, a division of Recognition Equipment, Inc., acquired the HP AIMS business from Hewlett-Packard.

IBM moved into the image market in 1988. Such slowness seems typical for IBM, which was a late entrant into other trends such as desktop publishing. The emphasis at IBM, of course, has always been on mainframe computers. Some analysts see image technology and IBM as having a natural affinity. Image processing, the argument goes, may increase the need for mainframe memory, power, and disk storage.

Today, IBM is promoting its Systems Application Architecture (SAA) Image and Record Management (IRM) System. IRM provides developers with utilities for image scanning, displaying, printing services, forms processing services (categorization, storage, and retrieval) and folder management services. A mainframe server sells for $250,000. It is targeted towards OS/2-based applications.

IBM's chief strengths are its installed base and history of customer satisfaction. The company has also selected numerous young hardware and software companies with good track records. Their products have met with good reception overall, especially in the finance, banking, and insurance arenas. While IBM has worked on increasing their ability to serve these markets, they have also been moving into new vertical markets, including government, health, pharmaceutical, and transportation. Marketing efforts also target three basic types of customers:

- Publishing systems producing high-quality output
- Occasional users who use documents not amenable to coding
- Users with operational systems that handle documents key to business operation, but which are not necessarily amenable to coding

The IBM SAA ImagePLus/400 family serves these targeted applications.

IBM also offers its customers applications and technologies linked to the Multiple Virtual Storage Enterprise Systems Architecture (MVS/ESA), Application System 400 (AS/400) and the Advanced Interactive Executive

(AIX) environments. IBM's software, on the other hand, uses the name ImagePlus. This imaging software was originally developed in cooperation with Citibank and the United Services Automobile Association (USAA). In fact, USAA actually developed the original software to keep track of the image file folders, which they licensed to IBM.

IBM is continuing to back its ImagePlus product family. Its release of ImagePlus for the AS/400 and MVS lines has been interpreted by analysts as proof that the firm intends to continue providing cost-effective imaging solutions for its installed base. In particular, the introduction of client-server capabilities to the MVS environment is intended to lower entry costs. The release allows for a site to set up a single, OS/2-based imaging system that later can be expanded to interact with the MVS mainframe. IBM also lowered the cost of entry for AS/400 users by changing its price structure. Pricing is no longer processor-based; instead, it is based on the number of users.

Eastman Kodak's goal is to provide vendors with support for the integration of its imaging products, emphasizing hybrid digital/microfilm storage solutions. For example, the Kodak ImageLink scanner/microimager 990S was developed to allow users to capture document images onto electronic media for processing and microfilm for archival storage. As further proof of Kodak's commitment to image, in 1992 it consolidated its core imaging business under the company's Office Imaging organization.

Unisys markets a strong line of UNIX server/PC-client LAN-based InfoImage products. Unisys targets three markets: public sector, financial, and commercial. Its InfoImage professional services, analysts say, have contributed to Unisys' recent turnaround. Entry level systems start at $245,000.

After NCR was acquired by AT&T, NCR launched its DMS (Document Managing System) imaging product and completed merging DMS with AT&T's imaging product line. The system is designed to provide a base hardware/software product suitable for applications, such as credit card processing and patient recordkeeping. A system for 20 workstation-enabled users starts at $300,000.

VARs

Meanwhile, large *value-added resellers (VARs)* concentrating on imaging have entered the marketplace to create systems that take advantage of the strengths of individual vendors. FileNet, for example, has sold solutions to American Airline and Diners Club. They have also supplied software and hardware solutions to IBM and AT&T.

Specific FileNet systems work with IBM PCs, Sun Microsystems, and Digital VAX workstations. All FileNet systems include at least some combination of the following:

- Document-entry stations
- Optical storage and retrieval libraries

- Image workstations
- Specialized software
- Print station software

FileNet's Image Management System (IMS) software is considered one of this business' key strengths, and includes a full range of office functions.

FileNet, however, has had is problems lately. In 1992, the company eliminated 10 to 15 percent of its 1,000 employees and suffered a $6.2 million loss. FileNet's vitality, analysts argue, will depend upon its ability to meet the competition from the growing number of vendors that emphasize lower-cost network solutions.

Sometimes, the line between VAR and vendor is blurred. IBM, for example, holds equity in Image Business Systems. Initially, IBS was a systems integrator, which made its own systems from third-party hardware and software. Now they create their own product line, providing the third-party hardware that IBM is notorious for supplying. Some analysts see IBS as an IBM VAR, but one whose own software provides the added value. Certainly, the semantics of the imaging industry are getting trickier.

Major players have also entered the imaging arena in recent years. Apple Computer made its first appearance at an AIIM show in 1991. The company demonstrated Macintosh-based image management systems along with 17 of its solutions partners. For example, Blueridge Technologies showed Optix Pilot, a Macintosh-based imaging system capable of indexing more than a million documents.

Thanks in part to the economic clout of these new vendors, the electronic imaging business has been growing exponentially in recent years. The Association for Information and Image Management found in a survey conducted by Temple, Barker and Sloane that the industry was having a five-year compound annual growth rate of 54 percent, and would reach the $6.8 billion mark in 1993. The AIIM annual trade show has seen attendance skyrocket from 12,000 in 1986 to 33,500 in 1991.

By 1990, EIM had almost caught up with micrographics in revenues. AIIM's survey of 560 members accounting for nearly the entire industry found that EIM revenues were $1.3 billion, compared to revenues of $1.4 billion for other micrographic products. (An additional $761 million came from services and $70 million from software.) Other important findings from the landmark survey included the following:

- Cost is a very important consideration in making imaging purchase decisions.
- A technology needs to prove itself before end users will implement it.
- Micrographics will remain a viable imaging technology over the next decade.
- End users need to be better educated on the range of imaging alternatives.

Today, electronic imaging, many analysts believe, actually has the upper hand. In March 1993, Datapro declared that "During 1992, the imaging market shifted from one dominated by micrographics to one dominated by electronic data processing."

Workflow

As discussed in detail in chapter 6, one of key technological developments in the computer in general has been the development of *LANs (local area networks)*. A LAN arises when you link computers together to share resources. One of the key advantages of a networked system is that it is capable of running more advanced applications than its standalone counterpart. LANs allow users to route and track correspondence, facilitating the most sophisticated use of image systems, *workflow* applications. Thanks to workflow applications, an organization can use a LAN to immerse itself in an electronic environment, consolidating and automating all phases of the business process. For example, in the case of an insurance company, an EIM service could locate field reports, find photographs, and locate past records, including billing histories and legal transcripts—tasks that might have been done in the past by many people.

Workflow automation has become a critical feature of imaging systems. In pre-computer offices, workflow was traced from the in-basket on one desk to the out-basket on another. Computers can, of course, move information from workstation to workstation. However, the enthusiasm for workflow goes way beyond the fact that it replaces the old "in-basket" with a queue of documents. Because workflow automates completion of each step of the process as the information is moved from application to application, workers are more productive.

Any increase in worker productivity in the office environment can be viewed as "revolutionary." While industrial productivity has increased 100 percent in the past 25 years in the United States, office productivity has increased only by two to five percent during that time. As Herbert A. Edelstein said in "Imaging Shifts Emphasis To Workflow Management": "Corporate America's investment in personal computers has done little to improve white-collar productivity. But there is still hope. Modern imaging technology allows a bitmapped image of document to be captured, stored, and made available within a database. The image can then be retrieved and distributed for processing . . .. The biggest payoff comes from using images to manage workflow."

The Veterans Administration, for example, recognizes the importance of workflow management. The agency hired American Management Systems (AMS) to set up a system to process claims benefits under the new GI Education bill. One of the cornerstone problems for the system was that it had to scan the 1,500 documents it receives every day. Each document then had to be indexed by claim number, form type, and scan date. All documents for the same case must then be grouped together.

Today when a case is opened, it is routed to a person in one of three divisions—adjudication, finance, and veteran services. Each person still has his or her own queue of documents to confront every day. However, now the documents reside as images that can be retrieved for processing. Workers call up other needed information from a mainframe database via their workstation windows. After a step is completed, each case is moved along the appropriate logical queue without any physical transfer. If a document needs further processing, it can be sent back to the workstation that handled it previously.

A key advantage of this system is that case flow monitoring is now easy. Supervisors can make sure high-priority cases are handled first. The elimination of transfer time also speeds up the process as a whole.

American Express and Rapid Growth

The Nolan Norton Institute said in a report in 1990 fewer than 2.5 percent of the companies surveyed had incorporated imaging technology, and 14 percent were in the pilot stage. In other words, the technology has evolved to the point that electronic imaging has moved into the domain of ". . . average company. It's no longer avant garde." This report also warned companies considering EIM at the beginning of the decade, "If you are not in the pilot test, preparing to ramp up, you are behind the leaders in your industry."

The 2.5 percent of companies who have managed to implement electronic imaging, however, have created a track record that many others will want to use to their advantage. American Express has been recognized as one of the pioneers in using electronic imaging as a way of increasing efficiency and as a competitive tool.

"By the mid-1980s, American Express was swimming in a sea of paper," Steve Madden explained in "American Express: Giving Billing The Right Image," *(Inform*, Fall 1990), "But there was a twist. This was no ordinary sea of memos, faxes, and reports. This was a vast Pacific-sized expanse of 3¾-by-5-inch blue and white slips."

The slips, of course, were American Express charge receipts. They poured into the company's Arizona and Florida billing centers at the rate of almost two million per day. Paper receipts were growing at the rate of three percent per year.

The company was using a process known as "country club billing," whereby the actual receipt was returned to the customer after processing. For this to happen, all of the charge slips that people signed when they made a purchase ended up coming back to American Express where they could be sorted down to the individual card level. The sorting process was messy. Operators keyed in charge amounts by hand, causing numerous errors. Also, the process was slow.

At the same time, as Cliff Dodd, American Express's senior vice president of billing and payment services explained to Madden, "We wanted to speed things up because time is money to us, as it is to anyone in the credit business. We pay the merchants when we receive the charge receipts, but we have to wait for the cardmember to pay us back before we realize our profit. If we could reduce the interest expense, we could realize a substantial savings."

American Express turned to TRW and Xerox to design a solution. The result was "enhanced country club" billing. This image-based system has gained the reputation as the most advanced in the world. When a cardmember makes a purchase, the merchant runs the card through a scanner and enters the amount of purchase via a keyboard. All information is transmitted to a main computer. Customers still sign their bills. At billing time, American Express cardmembers now get laser-printed copies of the slips they signed when they made their purchases rather than a pile of hard-if-not-impossible-to-read carbons. A single slip will contain as many as six or eight bills.

The project, launched in March 1985 and completed in July 1988, has reduced processing time for a credit card from four working days to 1.5 days. More than a convenience, American Express's management treated this process as a strategic advantage—a way of standing out. When customers decide which credit card to use, they tend to focus on the one that is handy or paid up. The new billing twist might have tilted some customers' decisions slightly in favor of American Express. Of course, any change also risks a customer backlash. A survey of cardmembers six months after the new system began, however, found that 92 percent said they liked enhanced country club billing better than the old system.

"The bill also gives us a whole new marketing platform," Dodd says. "There's a lot more room in the envelope now that all those receipts aren't in there. So we will be able to use the space to promote the benefits of the card and other services."

In addition, the company also claims that increased productivity and reduced cost of operation have resulted in more control over pricing. Completely documenting transactions, the company says, makes government contract compliance easier. Customer satisfaction may also have increased; some report that the monthly images are easier to read than the normal sales slips.

The Evolution Continues

In the late 1980s, a new class of imaging products started to enter the market. Unlike their simplistic predecessors, these systems provide sophisticated management for the flow of document images, as well as methods for storing and retrieving electronic documents. The new imaging system identifies and uses defining data fields within the document images. In other words, the computer knows how to separate what needs to be processed from data that doesn't. It can recognize the difference between "wheat" and "chaff." These newer, more discriminating systems fit nicely with traditional data processing systems, which essentially have the same focus, separating applications programming from the data the programming will address; a computer will process numbers on a Lotus worksheet or text written in WordPerfect, and leave the basic programming intact.

In fact, some industry gurus are predicting that electronic imaging will soon become "just another data type." The future, they say, will have the capturing of data, text, graphics, image, video, and voice all linked together. New systems will process information regardless of its source, type, or destination. While this might be true, no one can reasonably suggest that electronic imaging as a field will become obsolete. Instead, the field will broaden in the long run to keep pace with the times. Nor should any manager refrain from incorporating electronic imaging into long-range planning because of inevitable evolution. Indeed, good planning takes evolution into account.

Surprisingly, even today, you still find pockets of resistance to the very idea of electronic imaging. As recently as April 1992, *Inform* ran an article with the provocative title, "Acceptance: Imaging's Biggest Hurdle." The article, by Price Waterhouse's A. Brinkley Trammell, opens with a 1513 quote by Niccolo Machiavelli: "It must be considered that there is nothing more difficult to carry out, nor more doubtful of success, nor more dangerous to handle, than to initiate a new system. For the reformer has enemies in all those who profit by the old system, and only lukewarm defenders in all those who would profit by the new."

No one ever accused Machiavelli of being a computer nut. The stormy relationship between Apple and IBM was never mentioned in any of his writings. Trammell, however, used the cynical political maestro's observation to make the point that "it is human nature to resist almost any kind of change." Fortunately, he ends his article on an "up" note, pointing out that "our imaging systems will eventually be as common as personal computers and telephones. It will be easy for employees to accept imaging when they see it everywhere they go."

Of course, what we are seeing here is history repeating itself. Starting in the 1940s, computers themselves were the targets of various attacks. Science fiction writers created a whole genre of horror stories based loosely on the theme "the computer that ate civilization." Some even predicted that computers would soon replace the human brain. Meanwhile, turf-protecting unions forced strong anti-automation clauses into union-management labor contracts. Looking at the IBM clone or Mac on your desk today used to store recipes, play Tetris, compose songs, or whatever, it's hard to understand what all the hysteria was about.

Mastering the art of electronic imaging does require an adjustment. No one will deny that. Middle managers, especially, sometimes are visibly shaken at the prospect of a "significant strategic reorganization." They fear that their departmental fiefdom will disappear under the trampling hooves of progress. Or, even if their area of influence survives, they worry about facing an organization in which they will no longer know how to manage, because of fundamental changes in function, the specific tasks of employees, the tools of their trade, and the evaluation criteria for individual employees. After all, many of these middle managers have worked long and hard to set up a shop that works smoothly. Although these fears tend to be exaggerated, the fact remains that an electronic imaging system is doomed from the start without the support of middle management.

Day-to-day electronic imaging, however, tends to be the realm of clerical personnel. Highly paid professionals might oversee the process, but clerks understand what has to go on. They are the main source of practical electronic imaging expertise. Many physicians, fresh out of medical school and crammed full of theory, end up asking their nurses for advice on how to take care of "real live" patients with specific problems. Similarly, anyone interested in solving real problems with the flow of information and documentation around an organization would be wise to consult clerical employees.

The good news is that when questioned whether they would want to go back to their old ways, workers with new image systems overwhelmingly respond, "No!" Computers take over the drudgery of pushing papers around. This frees up more time for the typical employee to use his or her skills over a broader range or to concentrate more on things he or she likes to do.

From a purely technical point of view, electronic imaging is ideal. These systems allow for the automatic distribution of documents according to a predefined criteria, with the highest-priority documents always at the top of the in-basket. No document will end up stuck at the bottom of that basket. Electronic imaging could even be said to promote a sort of "workplace democracy," where everyone has access to whatever information he or she needs.

The problem, of course, is that increased efficiency sometimes leads to the elimination of certain tasks. There is no way of dancing around that harsh reality. The good news, however, is that the elimination of tasks does not have to mean the elimination of jobs. Job descriptions might become obsolete very easily, but good companies recognize the importance of preserving their "human capital."

Managers meanwhile are asking themselves questions about just how far they want their organization to go with this technology. Typical managerial questions emphasizing deployment strategies include these:

- Would a LAN or standalone system work better?
- Should systems be departmentally based or cover an entire enterprise?
- Should systems be used in workflow environments automating all steps in a business operation, or are the file and retrieve systems more appropriate?

Searching for the "right" vendor these days might also prove more difficult for companies. The imaging world is embracing a trend in which competitors who otherwise compete head-on realize competitive advantages through strategic alliances. These alliances are made possible through the sharing of technologies. Because of the volume and variety of technologies critical to imaging, almost no one is going it alone.

IBM, for example, is aligned with numerous small software firms that tailor IBM imaging systems to the specific needs of vertical markets. IBM gives these companies marketing clout. The small companies, in turn, provide IBM with extensive knowledge of particular markets that would otherwise be more difficult to acquire.

Larger companies can also cooperate. Digital Equipment, Unisys, and Wang have all made alliances. Both small and large vendors now play the role of system integrator, purchasing scanners, terminals, drives, and other parts through various OEMs. For example, Canon makes the driver for Hewlett-Packard printers.

How Much Does It Cost?

Let's face it. "What does it cost?" is one of the most crucial and perhaps difficult questions potential users ask when considering the implementation of imaging. In 1990 the Nolan Norton Institute presented a white paper at the annual AIIM conference estimating that a 20-workstation pilot program with hardware, operating software, and supplies would total $910,000. They offered a basic figure of $30,000 per workstation. This figure, however, does not include the costs of training or converting essential microform and paper documents. These costs can add more than $20,000 per user year. Other important added cost for an imaging system can include the increased expenditures involved in moving large blocks of information over remote networks. The Commonwealth of Virginia also researched electronic imaging and came up with an average cost of $50,000 per workstation for hardware, software, minimal paper conversion, and training. This same research estimated the payback period for research at three to five years. Noting that mid-range systems prices were falling, however, a 1993 Datapro report put the price tag for an average base system at $43,500.

Vendors and price-sensitive customers meanwhile have come up with numerous solutions to minimize cost problems. One way to reduce costs is to use lease-purchasing. The advantage of lease-purchasing plans are that affordable payments can be spread over a specified period of time. A well-developed plan can also provide support services at a discount. Some also include provisions for discounted upgrading.

Cooperative purchasing agreements are another way to lower the cost of an imaging system. Cash-poor state and local agencies can gain added leverage by making aggregate purchasing agreements with vendors. For this policy to work, however, cooperation and significant preparation are needed.

Governments and businesses can also cooperate by creating a joint service bureau. Public and private sector organizations can share costs through regional and geographic partnerships modeled after existing regional geographic information systems. Many of these systems team up public and private sector companies.

Imaging systems can also be used to generate revenue. For example, the Registry of Deeds for Middlesex, Massachusetts, provides value-added services, including online access to the Registry's database of deeds and titles, to title search companies and law firms. The Registry will search optical disc records from 1960 through the present and, if a record is found, fax it immediately to a customer. The office is able to raise over $100,000 annually this way.

If revenue generation is not an option, many systems can be justified through cost reductions. The Rochester, New York police department found that it could return two police officers to the street and eliminate four civilian jobs through attrition. The resulting reduction in labor costs would give the imaging system a six-year payback period.

Traditional measures of economic impact, however, might greatly underestimate the value of electronic imaging (or any other emerging technology). Indeed, there is a movement underway in business schools and progressive firms to develop a high-tech form of accounting. The idea is to bring old and perhaps outdated methods of accounting in line with technological realities. "If organizations are to survive in a complex technological environment, they have to break the return-on-investment (ROI) stranglehold. They must modify the process they use to account for technology investments," Thorton May of Nolan Norton writes *(Image Management*, November 1991).

Business leaders, indeed, have tried to work out a new way of evaluating programs. In 1990, Robert Kaplan, Harvard Business School professor, and David Norton, president of Nolan Norton, organized a research program sponsored by KPMG Peat Marwick to address this issue. Their goal was to document the theory that the methods companies use to allocate capital and evaluate people do not match the true determinants of economic success in the information age. In short, information drives industries today—but our evaluative mechanisms were developed in an era when land and capital were the key resources.

Businesses worldwide, according to May, are spending billions of dollars rebuilding their information infrastructure. Most of us have been trained to think of infrastructure in terms of roads and buildings. Instead, May uses the term to mean "shared information capabilities." He concludes: "One of the challenges facing information management in the 1990s will be to communicate to senior management which technologies have to be shared."

Because imaging systems can be costly to implement, May suggests that users and vendors adapt their contracts so that they agree upon a benefits-based payment schedule related to the delivery of value. The City of Chicago, for example, wanted to solve the problem of outstanding parking violations. The city did not want to accept traditional pricing arrangements. The vendor who won the contract proposed a risk-sharing, pay-for-performance contract stating that the vendor would receive 18 percent of all fines collected on tickets that were outstanding for less than a year and 26 percent of fines on tickets more than one year old.

May does not advocate the use of pilot projects. He argues that they take too much time—from 18 to 23 months. In addition, he says, they take the attention of senior management away from strategic issues, such as the ability of an organization to innovate and improve, the overall performance of the organization in areas of greatest customer interest, and high-level financial measures that are appropriate for both internal and external audiences. Instead of traditional evaluation techniques, May suggests a three-step process:

1. Generate online performance measurements of your known business needs.
2. Rapidly prototype emerging technologies.
3. Monitor emerging technologies.

Current Applications

Even elementary-school children are learning to work with electronic imaging. Pentax and Apple Computer have teamed up to create special systems for Educational Service District 101, serving schoolchildren in remote rural areas of Alaska and the Pacific Northwest. Satellite technology makes the program work.

"Our program will allow children to take advanced mathematics, science, technology, and language classes that previously were available only to students at larger urban schools," MARCOM specialist Cindi Holter of Pentax Technologies explained. "Until now, such programs were impossible due to small enrollments and the difficulty of providing timely feedback over the long distances between remote schools and cities. Televised education is as old as television itself, but now scanning technology and satellite links combine to allow quick response to assignments and tests that are necessary for advanced courses. We're also helping teachers by providing in-service career advancement services."

The program starts from ESD 101 headquarters in Spokane, Washington, where audio, video, and data transmissions emanate. These transmissions then are downlinked to cities scattered across the Pacific Northwest. Each downlink location has a Macintosh Classic computer, an IQ scanner, audiovisual equipment, an ImageWriter printer, modem, and customized software. Data is sent to ImageWriter, where it is printed for distribution to students. Assignments, notes, tests, and other documents are scanned on the IQ scan, then stored on the Classic, and transmitted back to ESD 101. Grades are posted via satellite, which also enables students to study together even when they're thousands of miles and two time zones away.

The State of Washington's Department of Wildlife, meanwhile, uses a Cognitronics character-recognition system to handle three million documents each year, including hunting and fishing licenses. The system reads handprinted numbers and checkboxes and can handle up to 4,000 forms in an hour. It runs on two IBM-compatible PCs. The system has reduced printing costs for the department by $100,000 a year because scanners have reduced the number of standardized forms.

The federal government is also an important force in the imaging industry. A 1991 survey by *Government Computer News* predicted that this segment's electronic imaging budgets will grow seven percent per year through fiscal 1995. The seven percent figure, however, might underestimate the extent of growth in this segment. Thirty-eight percent of all respondents said they had plans to implement an imaging system between 1992 and 1995, while another 30 percent were evaluating systems.

The average overall price tag for an installed imaging system, according to this survey, is $240,000. However, 21 percent say they spent between $25,000 and $99,000 and an almost equal number report spending more than $500,000. The most prominent applications used by federal agencies, according to the survey, are the following:

- Records management and retrieval
- Electronic publishing
- Engineering drawings
- Archiving

Other key findings of the survey include:

- 15 percent plan to invest in optical jukeboxes
- 15 percent expect to buy a CD-ROM system
- Standalone micrographics account for 33 percent of dollars spent
- Microcomputers are the most popular platform, named by 77 percent
- 60 percent run database management programs
- More than half say their setups include LANs

Even the Department of Defense is getting into the electronic imaging arena. In fact, *Government Computer News* found that this agency alone accounts for 27 percent of all federal imaging spending. Journalists joked about the irony of using the term "data capture" to describe the Defense Intelligence Agency's widely publicized scanning of 12 million pages seized from Iraq during the Gulf War. Work for this massive project took place in Kuwait City, the Saudi Arabian city of Dharan and DIA headquarters. DIA officials scanned truckloads of contents from 73 Iraqi filing cabinets at 120 pages per minute and 200 dots per inch on two Eastman Kodak Imagelink Scanner 900s. The scanners were linked to an IBM PS/2 Model 95 equipped with an IBM database manager and 8514 XGA monitor. In addition, officials used a special mobile scanning unit for airlift. It consisted of a PS/2 Model 75, a Bell & Howell CopiScan Series II scanner, and an optical drive.

The process reportedly took about a month to scan and index. The DIA worked with a team of 10 translators and two shifts of nine people each preparing the documents, as well as four teams of two scanning. The translators identified key words within the documents to facilitate indexing. All scanning was done in black and white. Header sheets were created to show date of capture, place found, and type of document. The teams keyed in information meanwhile to create a usable database.

Yes, it's no secret: the Central Intelligence Agency is using electronic imaging systems. Delta Tech has a low-profile contract to create a simple and fast image storage and retrieval system for the CIA, according to *Military and Aerospace Electronics* (April 1, 1991). The magazine reports that the system has sparked interest in other government agencies, including the Naturalization Service.

You might not think of universities as having the need for a competitive technological edge. However, University of Southern California associate dean Tom King points out that the development of an imaging system for his admissions department ended up giving his school an edge. King's article "Imaging Frees Admissions From Backlog" in *Government Technology* (May 1992) explains, "By getting our decisions out to the students sooner, our chances of having them choose USC are improved substantially."

USC's primary goals were to increase the speed of the application process and to decrease valuable storage space on the small campus located in Los Angeles. The college faced a Byzantine system of filing, copying, and handling applications through the evaluation process. All applications arrived at the undergraduate admissions offices, including graduate applications. From there, 60 percent of the applications went off to the Office for Graduate and International Admissions or to academic departments for admission decisions. The applications were color-coded for filing and then photocopied before they left those offices.

The process was slow, to say the least. Manual handling of these documents created eight to ten week backlogs during peak periods. An unknown number of documents were misfiled so that, according to King, "The staff spent a great deal of time trying to correct the problem."

USC decided to try a pilot project and turned to Wang Laboratories and Trident Systems for a pilot. In 1989, they purchased a system consisting of the following:

- A Wang VS image server
- Five document scanners
- An optical jukebox
- 37 image-capable PCs

Most of the equipment was housed at the USC Wang Processing Center. Two remote sites were set up, however, one for the Graduate and International Admissions office and another at the Registrar's office. They were then linked by fiberoptic cable.

Today the admissions process takes place in seven steps:

1. The mail is opened, sorted, coded, and batched.
2. Documents are scanned.
3. Pertinent information is entered into the University's central database using a Prime mainframe system. This database contains financial and academic data.
4. Documents in the database are indexed by document identification number with correct document type and the applicant identification number.
5. Quality assurance technologists check the indexing for accuracy.
6. Every evening, the database selects all student documents that are ready for submission.
7. The next day, counselors working at imaging workstations request and display records.

The results? Turnaround time during peak periods has been reduced an average of six weeks. A recent increase in admissions applications (30 percent) did not require additional staffing or affect turnaround time. USC also added a fax gateway to receive faxed applications from prospective students.

King concludes, "While much of the success with imaging in USC's admissions office can be attributed to the virtues of the technology and its ability to automate a slow manual system, a great deal of it has to do with our staff's ability to adapt and use imaging in a busy environment. Thanks to some good initial training from Wang, the admission staff quickly became comfortable with the system, and has taken over training of others as they join the department. Imaging has taken the unnecessary stress out of the job by giving the staff the confidence that everything they need is right in front of them—on their desktops."

An expansion program at USC now promises to bring EIM to financial aid operations. The university has doubled the size of its Wang server, doubled the number of imaging workstations connected to the system, and quadrupled optical jukebox capabilities. This process involves the management of more than a million pages of historical documents before new ones can be filed.

Phoenix, the pioneering city

A growing number of civic-minded institutions are starting to realize the power of electronic imaging—thanks in part to a pioneering project recently completed in Phoenix, Arizona. The City of Phoenix is the eighth largest city in the United States, employing over 11,000 workers. Since its creation in 1881, the city has burgeoned to one million residents. You can imagine how many records that phenomenal growth produced. By 1988, in fact, the City Clerk department, the agency responsible for ordinances, resolutions, contracts, deeds, and requests for council actions, had accumulated more than two million pages of official records. On top if that, another 150,000 records were being heaped onto the existing pile every year.

Problems arose when the 38-year old Kardex system used to cross-index records "burst at the seams." Clerks drummed their fingers and tapped their feet, waiting for the system to work better. It didn't. Not surprisingly, studies had long recommended automating the system. Also, not surprising, funds were not available for such a project.

What followed was a monumental civic effort to revitalize the municipal records system under the banner of electronic imaging. The City Clerk department took its studies to the bond election committee, the first hurdle in Phoenix fund-raising. The committee recommended the idea of a bond levy to the city council. The council gave its approval to put the matter to a vote. Then it was up to the voters.

The levy passed. In April 1988, the citizens of Phoenix approved a $1.03 million bond request. But that was only the beginning of the electronic imaging process. Project manager Susan Antonini launched an investigative quest that started with demonstrations of equipment capabilities and included discussions with professional organizations and officials in other cities. Then she reached a decision: What the City of Phoenix really needed was an optical disk system for the storage and retrieval of all the official records.

Next came the Request for Proposal (RFP), the first step in the bureaucratic mating dance. The city hired Trident Systems, a consulting firm, just to handle the RFP's climb up the ladder of rejection, starting with the city's Information Master Planning committee, the watchdog group that reviewed the proposal for compliance with other city programs. The climb was long and tedious, but proponents understood that the process was necessary, just like the politicking involved in any other investment decision. They also reasoned that this process would "secure the involvement of top management"—an extra bonus. In the end, Wang Laboratories won the contract.

Even before equipment was purchased, offices had to be remodeled. Decision trees bloomed like sagebrush throughout the department. Cooling requirements, space needs, and weight tolerances all became critical issues. Even cabling was a challenge. Should they use coaxial or twisted-pair cables? What are coaxial cables anyway? Then, everything had to be checked against building codes. Nothing would have been more embarrassing than having a city office in violation of Phoenix's own building codes.

Meanwhile, the system itself was under construction. Trident's services were extended to include preparation of the database design specification. The city, Wang, and the consultant all agreed that the database would be the key to the systems' success or failure; everyone agreed to review the specs (specifications) during the design phase.

The hardware was installed in August 1989. The database, which only needed minor changes, was ready in November. A full training program, meanwhile, was underway. Records clerks started by attending general office automation classes to get comfortable with the idea of using workstations. The first imaging terminals arrived after the staff had already learned the automation functions. Off-site vendors took care of the next training phases. Finally, full production began in January 1990. Wang kept a trainer on-site to field questions and solve any lingering problems.

The final results of the two-month effort? The average retrieval rate decreased from five minutes to 30 seconds. In other words, efficiency improved tenfold. New services were added. Citizens can now obtain documents simply by faxing their requests. The city faxes the documents right back from the imaging system. In addition, the space savings from eliminating the miles and miles of files was considered "enormous." The entire system is comprised of 12 image-capable workstations, one jukebox, two scanners, and two printers. It is connected to the city's Office Automation Support Information System (OASIS), linking over 400 PCs and terminals throughout the city. The facsimile gateway allows documents to be faxed directly to any fax machine in the world. Of course, it can receive any fax as well.

This new system also allows for word processing, electronic mail, and electronic scheduling. Special built-in features also offer time-saving services. A function known as Agenda Results allows staffers to call up the outcome of city council meetings on their workstations the morning after the meeting. Clerks can prepare bid specifications and other documents that had previously been done on woefully inefficient typewriters. In many cases, accessing this material is done with the simple pressing of a few buttons to recall it from a glossary.

"Like most government agencies, the City of Phoenix has a mission to provide service to the public," Antonini summed up in the March 1992 *Inform* article "Rebirth of a Records System." "Such service was the foremost goal of this project. Phoenix now runs a records management program appropriate for the eighth largest city in the United States."

Palm Beach County

Palm Beach County also used electronic imaging to solve a records problem. In Florida, registering vehicles is the responsibility of county tax collectors. The problem was that vehicle registration numbers were exploding. In 1981, the county's rolls listed 536,981 vehicles. That figure jumped to 900,111 in 1991. As a result, the tax collector's office filled three large elevator file systems and was considering purchasing another to try and cope with the problem. Meanwhile, the tasks of filing, sorting, and retrieving those records had become a nightmare. Misfiles and damaged documents were everywhere.

The department wanted a system capable of solving these problems. Other goals included quicker access to records, a reduction in storage space, and the ability to manage growth without adding staff. Legal issues were another important consideration; the county needed to have a record system maintaining the same legal standing as original documents.

The solution was microfilm. In early 1989, Palm Beach County installed KMS software on its mainframe computer. Workers began indexing and retrieving records with Kodak IMT-350 and IMT-200 microimage terminals. They are now keeping pace with the 8,000 that six vehicle-registration offices collect each day, as well as attacking the backlog. Meanwhile, other workers are indexing documents on the computers, sometimes as many as 10,000 each day. All of this takes place against the backdrop of a high-pitched moan as the office paper shredder chops up the pile of aging paper documents.

The department is looking for smaller quarters now that the old files are disappearing. They also can promise fast document searches—as little as 30 seconds per search. No jobs have been lost, but neither does anyone worry about soaring personnel costs. They also have an impressive way of double-checking for errors. When records come in, clerks enter information directly into the office's mainframe computer. This information is then checked against the microfilm.

The IRS

The Internal Revenue Service is a paper-pushing nightmare. With an average staff of 120,000 working at 10 regional service centers and a host of district offices, any normal agency would have problems. The IRS, however, is no ordinary agency. It receives 40 million letters annually and sends out 80 million refunds every year, as well as eight million notices to taxpayers who have made errors on their returns or forgotten to sign them. All of this paperwork, of course, is subordinate to the agency's goal of collecting over $1 trillion in taxes annually.

The IRS has also been a technological nightmare. "We have been running our business essentially the same way for more than 25 years," Fred Goldberg,

IRS commissioner, lamented before the Congressional Subcommittee for Commerce, Consumer, and Monetary Affairs. "The need [for modernization] is compelling." Indeed, the IRS has admitted that their outmoded system accounts for six to seven million unnecessary contacts with taxpayers each year. Responses to taxpayer questions sometimes take as long as 30 days. The subcommittee agreed with the IRS, calling the current IRS system "error prone" and "labor-intensive."

Fortunately, the situation appears to be on the mend. The IRS has been image-technology hungry for years. They even tried to modernize their computer systems between 1968 and 1978. Their plans, however, ran up against congressional concern about cost, privacy issues, and taxpayer information security questions. Another attempt fizzled in 1986. Known as the Tax System Redesign, it suffered from insufficient technical expertise and floundered in a sea of management changes.

The IRS is now in the middle of the latest attempt at tax systems modernization. The outlook is very favorable. Few organizations have spent as much time experimenting with image prototypes or conducting extensive studies as the IRS. For example, in 1985 the IRS set up a prototype image storage and retrieval system at its Fresno, California facility. (The program, by the way, was judged a success and cut the time needed to obtain an individual's tax return from six days to overnight.) Certainly, modernization can be cost-justified by hopes of an unusually high return on investment. Any technology that would help the IRS collect the roughly $112 billion of uncollected tax liabilities each year deserves serious consideration.

This modernization program, an $8.3 billion, 12-year project, is often recognized as the biggest automation project ever. The program calls for extensive use of image character recognition technology to convert the more than 200 million tax returns the IRS receives annually. The IRS claims that the program will

- Ensure that more taxpayers pay the right amount of taxes thanks to improved case reporting and auditing procedures
- Generate billions of dollars in cost savings and additional revenue
- Enable the IRS to respond to taxpayer inquiries in a timely manner

The ultimate goal is to create a paperless office. As you can imagine, the process is extremely complex. Planners are breaking the 52 separate modernization projects needed to complete this task into four separate groups:

- Applications
- Procurements
- Infrastructure
- Management

The applications group alone is responsible for 23 of these projects, including:

- Dealing with taxpayers
- Cash and funds management
- Case management information system
- Data storage and retrieval

The cash and funds management program will allow for electronic disbursement of funds and deposits. The case management information system will take advantage of the online databases created at the service centers. In many cases, IRS employees will be able to solve problems online. Many taxpayer problems that now require several face-to-face meetings will be able to be solved in a single meeting. Nineteen of twenty taxpayers, they say, will be able to resolve tax matters by dealing with a single IRS employee. Almost all of those annoying computer-generated enforcement notices that, under the current system, are mailed out after a taxpayer has responded to an earlier notice also promise to be eliminated.

The biggest obstacle to all of this seems to be the capturing and input of paper returns. ICR technology uses algorithms and artificial intelligence techniques to read typed or handwritten characters on tax forms. The agency believes that these technologies will enable them to input somewhere between 60 and 90 percent of all data on forms. However, as many critics point out, the technology has not been demonstrated in an environment as large and complex as the IRS.

Manufacturing

Of all the industrial segments, manufacturing seems to be one of the slowest to adopt imaging. However, imaging is finding its niche there. For example, the pharmaceutical company Burroughs Wellcome has implemented an electronic imaging system. Every day, its clinical data department receives approximately 2,500 pages of clinical data submitted by physicians and others. Quick processing is crucial. The company must be able to respond immediately if test results indicate a trend of adverse reactions to a drug compound. In addition, incomplete or incorrect documentation of trial results could delay FDA approval, a setback that could cost the company hundreds of thousands of dollars.

Burroughs Wellcome decided to upgrade its current system, putting a high priority on the idea of making appropriate use of its existing systems, which included 1,000 PS/2 workstations and over 2,000 other PCs. It also decided to employ an Ethernet LAN for its high-volume imaging traffic. A FileNet system closely integrated with Burroughs Wellcome's existing mainframe and PC workstations was installed in February 1990. The complicated workflow process was automated using FileNet's WorkFlo Business System, an integrated package of document-processing software modules and other tools.

The new system was set up so that each page of an incoming case report is indexed and verified. System controls make sure that no pages are lost or misfiled. Workflow is set up so that files are automatically routed to professional editors who review the documents for thoroughness, then completeness. Medically trained staff, known as monitors, also may be consulted to validate data.

Subsequent adjustments must be filed using special change authorization forms. The imaging system creates these forms, which are then mailed to physicians as needed. The system also keeps track of the timing of the release of forms. This feature enables Borroughs Wellcome to maintain an accurate schedule of update requests. The system works because once the change authorizations arrive, the system electronically "paper clips" the forms to the original pages. This provides an audit trail of all changes made to documentation.

In the end, for Burroughs Wellcome, an integrated image solution offers the following benefits:

- Improved processing efficiency
- Increased information accuracy
- Increased professional productivity

Goshen Rubber is another example of a manufacturing company making good use of electronic imaging. Goshen Rubber manufactures seals and gaskets that it markets primarily to "tier two" auto industry suppliers. Every month, the organization uses imaging to track and process roughly 600 price quotes averaging five or six pages each.

Goshen Rubber has been using Wang Laboratories equipment for more than 10 years. Much of the company's Wang Integrated Image Systems (WITS) has been installed on existing hardware, including the company's Wang VS 1000 Model 50 mainframe. With the implementation of imaging, Goshen has added a two-drive optical jukebox and 33 image-capable terminals to its hardware.

Jim Holsopple, Goshen's manager of data processing, points out that imaging offers Goshen the following advantages:

- The average time to process a quote is cut from 15 to eight days.
- Information no longer gets misplaced.
- Accountability has improved.
- Goshen can give a potential customer the exact status of a quote within minutes.

Imaging as Fire Protection

Normally, no one thinks about electronic imaging as a form of fire protection; it is not promoted as such. However, Comcast Corporation proved that imaging does have this additional benefit. The company, a franchisee of Muzak as well as a communications and cable firm, thought that it avoided problems when fire gutted an office building in February 1991 but left its entire floor unscathed.

The bad news, however, was that toxic chemicals, including dioxins and PCBs generated by the building fire, contaminated the entire office. Comcast had lost 18 to 20 million documents, including corporate papers, original purchase agreements, and loan documents. City officials barred them from removing anything from the premises.

What could management do to retrieve at least the essential documents? They turned to electronic imaging as a solution. Genesis Data Technologies designed a system for them based on eight DECstation 320 platforms running over an Ethernet network.

In fact, Comcast used four networks. A video network used by Comcast management helped them select which documents to scan and which to throw away. Two networks shipped selected images from contaminated offices to an adjoining building. Another network provided a direct PC-to-PC feed for retrieving information on floppies and hard disks.

Obviously, the logistics were complicated. Scanning technicians had to wear protective clothing just to enter the contaminated floors. Still, Comcast retrieved 1.25 million documents, including all of those that were crucial to operations.

Does Your Organization Need an Electronic Imaging System?

Not every organization needs electronic imaging. On the other hand, many organizations could benefit from electronic imaging systems that are enhanced by workflow assistance. Sometimes, traditional electronic imaging techniques are too simple."If the majority of the documents to be filed can be originated electronically, a traditional data processing system can often suffice, especially if they can generate images from text," George Hall, a Tucson, Arizona community college professor explains. "For example, if a large city wants to reduce all of its traffic tickets to images, it might also consider using other technology to record the ticket digitally on the spot and have it transmitted to the proper agency by cellular phone or by some other network." This approach, also known as electronic data exchange (EDI), however, might end up being more expensive or error-prone than a traditional electronic imaging system.

Here are five potential warning signs that your organization might need electronic imaging to solve workflow problems. Answering "yes" to any of these questions might mean your organization needs electronic imaging:

1. Does your organization face a large caseload, a constant flow of documents, each day?
2. Is the number of documents your organization must handle growing?

3. Are you having trouble keeping or revising your documents?
4. Do you enter large amounts of paper-based information?
5. Are your documents needed by many people in several locations?

Another way to approach the question is to review records management policies. Once again, "yes" answers to any of these four questions makes your organization a likely candidate for electronic imaging:

1. Do your documents require multiple revisions?
2. Do you spend a lot of money storing documents?
3. Does your organization do much of its clerical work processing documents "by hand"?
4. Does your organization process a large number of documents?

Ultimately, different types of organizations have different expectations for electronic imaging. The 1990 AIIM survey found that most service organizations are interested in electronic imaging to improve customer service. This segment, including insurance, healthcare, banking, finance, advertising, and publishing, is by far the imaging industry's biggest spender. Government agencies emphasize the need to reduce storage space. Manufacturers want to decrease access time. This segment, although less paper-intensive than the others, has expressed a willingness to buy new products.

Summary

Image processing is here to stay. Many large companies and government agencies already consider electronic imaging an integral part of their day-to-day operations and an important strategic tool for increasing efficiency and keeping costs down. Others are giving the technologies serious consideration and are studying ways to implement it. Tumbling prices and the miniaturization trend guarantee that image processing will become increasingly affordable for smaller firms and the PC market. That's why it makes sense for every organization to learn about the technology—even if the effort turns out to produce a decision to postpone using the technology or even forgo using it altogether.

Chapter

2

Imaging Concepts and Benefits: Management's Perspective

Electronic image management is a method, system, and technology for converting and processing paper documents and/or their microfilm copies to digital electronic information. The information on the documents may be text or graphics, handwritten or printed. EIM systems have proved advantageous to all areas of governments, business, and industry.

For thousands of years information has been recorded on paper; for the past 60 years a considerable amount of information has been recorded on microfilm; for 30 years much information has been recorded on magnetic media (tape, disks, etc.). In each case, the new medium provided an additional method of storing information and did not replace the earlier media. EIM systems generally use optical disks as the dominant storage medium. (However, it would be a mistake to conclude that optical disks are the only component of an EIM system.) Optical disks provide an additional storage option to paper, microfilm, and magnetic media, all continuing to be beneficial as tools helping people and organizations manage information.

EIM began with the use of computers. Electronic information processing came into widespread use in the 1960s, but mainly for "crunching numbers." In the 1970s, electronic information processing was expanded to include text. In the 1980s, graphics were added. In the 1990s, images (documents), voice, graphics, text, and numbers can all be electronically processed (see Figure 2.1). Managers can no longer afford to think of graphics, text, and voice as discrete entities. They are fully integrated in today's computer environment.

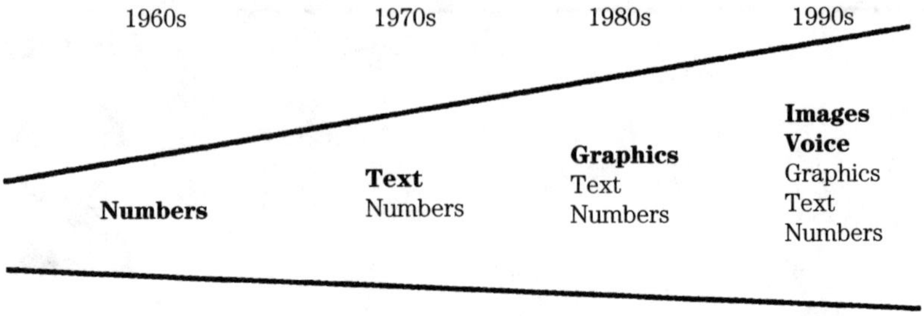

Figure 2.1 History of electronic information processing.

EIM, at its most basic level, is the application of computer technologies to document handling, primarily handling those documents historically that have been created on paper. As Chapter 1 points out, paper is not easy to manage. The biggest bottleneck in the operation of most businesses, governments, and institutions today is paper document-handling. Items including checks, correspondence, invoices, purchase orders, engineering drawings, and forms of all types are, for the most part, processed by hand.

Even the paperwork problem, however, is comparatively superficial. The bigger problem facing managers today is how to increase the productivity and output quality of white-collar staff (such as engineers, bankers, secretaries, lawyers and managers). To accomplish this, the entire business process of an organization must be reviewed and usually changed. EIM's main advantage for managers is that it can be the foundation for improving an organization's entire business process. EIM systems can provide the following advantages and benefits to managers:

- Reduces clerical support, processing time, and storage space
- Improves efficiency, customer service, document control, and decision-making
- Increases competitive advantage

Manufacturing plants have been retooled over the years to stay competitive. Who would want to buy a steel plant, for example, that is not state-of-the art? EIM brings this same philosophy of technology-based efficiency to the office. To ignore EIM is to mire an office in technological obsolescence.

EIM Activities

The key word in EIM is *management*. There are many managerial aspects to studying, planning for, and implementing an EIM system. These broad categories can be broken down further. For example, studying can be broken down into survey work and a feasibility study. Each of those, in turn, can be broken down into separate tasks. Any manager working in EIM probably has to address at least the following tasks:

- Survey
  —Establish a task force
  —Inventory records
  —Review existing systems
  —Determine user requirements
  —Establish organizational requirements
  —Review literature
  —Visit other sites
- Feasibility study
  —Identify applications
  —Size the project
  —Make cost/benefit analysis
  —Consider trade-offs
  —Develop a timetable
  —Establish technical requirements
  —Consider legal issues
  —Consider standards
  —Obtain management approval
- Systems development
  —Establish a strategy
  —Develop a detailed design
  —Run a pilot test
  —Prepare RFP/RFQ
  —Develop a disaster plan
- Implementation
  —Do the conversion, if required
  —Establish vendor relationships
  —Install the system
  —Promulgate a disaster plan
  —Train operators and users
  —Review and evaluate

Managing EIM Information

As Chapter 1 points out, EIM can be defined as an automated system to store, retrieve, transmit, process, and manage documents. However, on another level, EIM can focus on the technology of the information being processed. EIM systems preserve the visual and spatial characteristics and appearance of the original paper document, including handwritten and

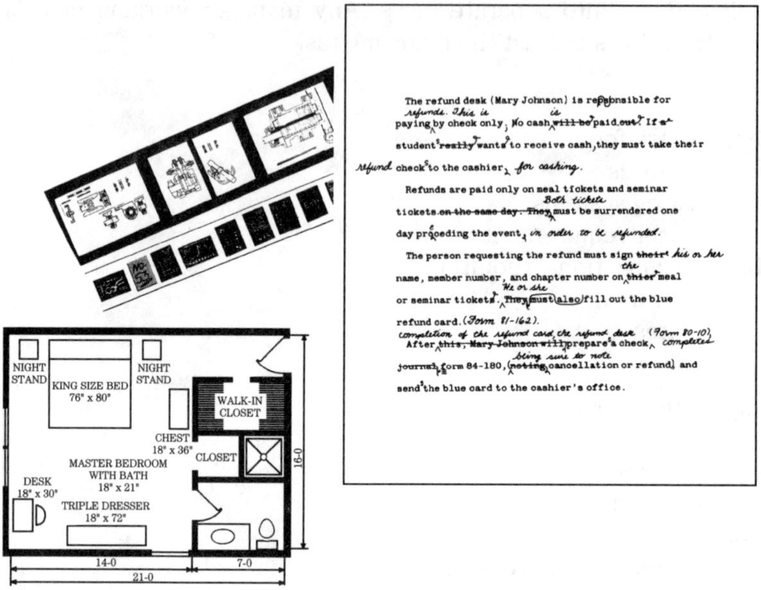

Figure 2.2 Samples of documents that can be handled by electronic imaging.

printed text, graphics, photographs, and engineering drawings. The EIM information universe deals with documents in three forms:

- *Analog.* Human-readable information via paper, microfilm, and other media. (Figure 2.2)
- *ASCII/EBCDIC.* Binary digital coded representations of information for computers on magnetic tape and disks (Figure 2.3)
- *Bitmapped/raster.* Binary digital representations of information for electronic imaging systems on optical disks and facsimile systems (Figure 2.4)

Understanding the difference between human-readable and coded forms of information is easy. However, the differences between ASCII/EBCDIC and bitmapped/raster, when both are binary digital coded representations, are not clear to most people. It is this difference that distinguishes EIM from other electronic information systems.

Character-based digitization

In computer systems, word processing systems, and OCR systems, every letter of the alphabet or number is represented by a combination of eight bits (eight zeros and ones). These standard coding schemes are called ASCII and EBCDIC. In these systems, any time a character is typed at the keyboard or read in through OCR devices, the computer recognizes the spe-

Symbol	ASCII	EBCDIC	Symbol	ASCII	EBCDIC
(space)	0100000	01000000	N	1001110	11010101
!	0100001	01011010	O	1001111	11010110
"	0100010	01111111	P	1010000	11010111
#	0100011	01111011	Q	1010001	11011000
$	0100100	01011011	R	1010010	11011001
%	0100101	01101100	S	1010011	11100010
&	0100110	01010000	T	1010100	11100011
'	0100111	01111101	U	1010101	11100100
(	0101000	01001101	V	1010110	11100101
)	0101001	01011101	W	1010111	11100110
*	0101010	01011100	X	1011000	11100111
+	0101011	01001110	Y	1011001	11101000
,	0101100	01101011	Z	1011010	11101001
-	0101101	01100000	[	1011011	01001010
.	0101110	01001011	/	1011100	
/	0101111	01100001	]	1011101	01011010
0	0110000	11110000	^	1011110	
1	0110001	11110001	—	1011111	
2	0110010	11110010	a	1100001	10000001
3	0110011	11110011	b	1100010	10000010
4	0110100	11110100	c	1100011	10000011
5	0110101	11110101	d	1100100	10000100
6	0110110	11110110	e	1100101	10000101
7	0110111	11110111	f	1100110	10000110
8	0111000	11111000	g	1100111	10000111
9	0111001	11111001	h	1101000	10001000
:	0111010	01111010	i	1101001	10001001
;	0111011	01011110	j	1101010	10010001
<	0111100	01001100	k	1101011	10010010
=	0111101	01111110	l	1101100	10010011
>	0111110	01101110	m	1101101	10010100
?	0111111	01101111	n	1101110	10010101
@	1000000	01111100	o	1101111	10010110
A	1000001	11000001	p	1110000	10010111
B	1000010	11000010	q	1110001	10011000
C	1000011	11000011	r	1110010	10011001
D	1000100	11000100	s	1110011	10100010
E	1000101	11000101	t	1110100	10100011
F	1000110	11000110	u	1110101	10100100
G	1000111	11000111	v	1110110	10100101
H	1001000	11001000	w	1110111	10100110
I	1001001	11001001	x	1111000	10100111
J	1001010	11010001	y	1111001	10101000
K	1001011	11010010	z	1111010	10101001
L	1001100	11010011	{	1111011	
M	1001101	11010100	}	1111101	

Note: Blanks in EBCDIC column mean symbol is not represented in EBCDIC code.

Figure 2.3 ASCII/EBCDIC computer code.

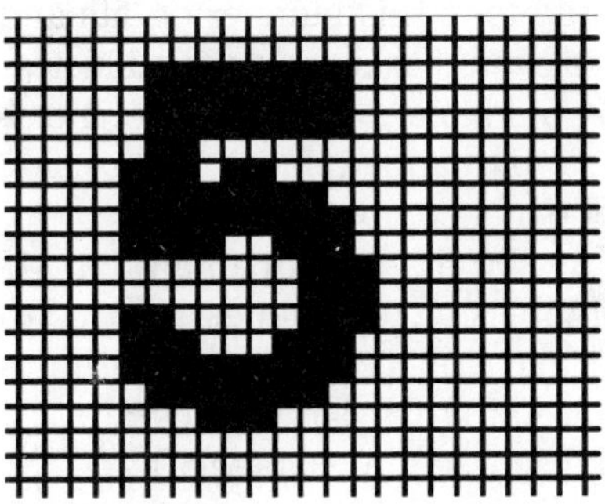

Figure 2.4 A bitmapped image.

cific character by its unique code. All machine manipulation, transmission, etc., is done by the code for each letter or number. The information is handled as intelligent text. When the characters are displayed or printed out, the eight-bit code is converted back to human-readable form.

A document with 400 words using ASCII or EBCDIC code would take about 24,000 bits, which equals 3,000 bytes (3K). There are eight bits per byte.

Bitmapping

In electronic imaging systems such as those employing optical disks, the image on the document is raster scanned. This technique is called *bitmapping*. In this process, characters are not recognized; the scanner simply converts light and dark points on the page of either text or graphics into bits of digital information. The entire page of the document is treated as a rectangular array of points called picture elements or *pixels*. In this process, binary digital coding is used to represent the dark (black or near black) and light (white or near white) pixels. The pixels are stored and transmitted in digital raster format exactly as a picture is written on a TV monitor; that is, left to right and top to bottom.

Photographs and colored images can also be scanned with a more complex extension of this process. Adding more information bits per pixel for grayscale adds memory requirements (fewer pages per disk). For 16 levels of gray (four bits per pixel) at 300 dpi (dots per inch), 4.2 megabytes are required. For 256 levels of gray (eight bits), 8.4 megabytes are needed for a single 8½×11-inch page. This compares to one megabyte at 300 dpi for a two-level black and white page. Some black and white scanners have a feature called *dithering* which is an electronic screening of the image. This feature improves pictures scanned with only one bit per pixel, providing a coarse likeness of the picture. The information is preserved, but not the picture quality. Images cannot be successfully scaled-up (enlarged) using the dithering technique. Gray-scale scanners, on the other hand, permit scaling and sizing. They can capture and preserve continuous-tone material with acceptable to very good quality.

An 8½×11-inch document raster scanned (bitmapped) at 200 dpi (200 pixels horizontally and 200 pixels vertically) would take 3,740,000 bits, which equals 467,500 bytes (467K). It would take this number of bits and bytes regardless of the number of words, or even if there were line illustrations on the page. Without compression, it would take 150 times more memory to store a 400-word document on a bitmapped basis than using character-based digitization with ASCII code in a computer.

Because it takes almost 4,000,000 bits at 200 dpi to bitmap an 8½×11-inch page and 8,500,000 bits at 300 dpi, compression techniques are used to reduce the number of bits to represent the image on a page. An algorithm is used to eliminate coding much of the white area (or large black areas if there are any) on a page. (Compression is discussed in chapter 5.)

Electronic imaging may be thought of as a technique of taking electronic pictures of documents. This technique preserves the visual and spatial characteristics and appearance of the input document. Electronic images may be created from information on paper or microfilm. Although most electronic images are entered into the system by scanning, there are other techniques for bringing images into the system. Converting information from other electronic information systems, such as CAD/CAM systems, is possible.

Advantages of EIM

One of the great advantages of electronic imaging is its universality. Electronic imaging systems can capture, retrieve, and transmit documents with all types of information: handwriting, typing, diagrams, pictures, fingerprints, etc. These abilities are illustrated in Figure 2.5.

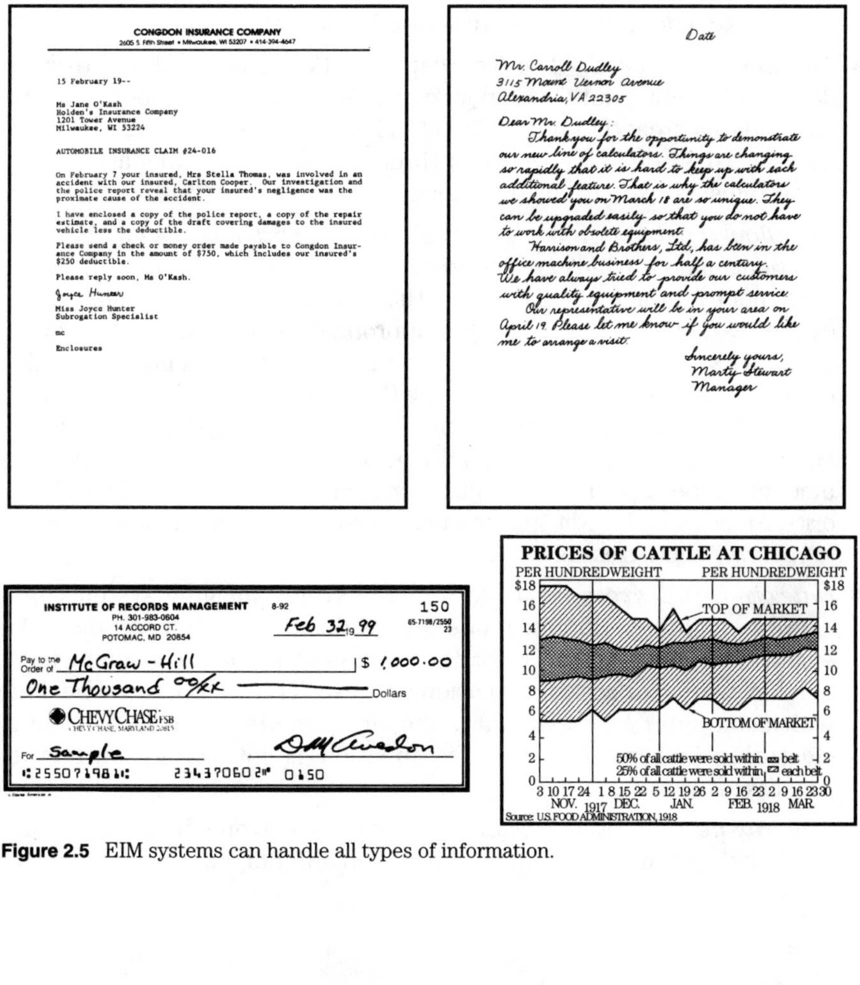

Figure 2.5 EIM systems can handle all types of information.

Many of the other benefits of EIM systems are most clear when you compare features to other technological systems. Electronic imaging systems, for example, can be compared to paper and microfilm. The advantages of EIM include the following:

- *Packing density.* The contents of 10 full four-drawer file cabinets (200,000 pages) can be put on one 12-inch optical disk. It would take 1,200 microfiche or forty 16mm reels of microfilm at a 24X reduction to store this amount of material. It would take 40 reels of magnetic tape, or 2,000 360K floppy disks, or 400 1.44M floppy disks to store this much. (This comparison is made with the images in bitmap form on both the optical and magnetic media.)
- *Very fast retrieval.* Using electronic imaging systems, any document from a file of hundreds of thousands of pages can be retrieved in a few seconds. From a file of millions of pages, retrieval should not exceed 30 seconds. With paper in a local file, five minutes would be a very good retrieval rate; from a remote site it could take hours. With microfilm it would take more time than optical disk but less than paper.
- *Transmission speed.* The fact that optical or magnetic disks store the document digitally rather than in analog form as it is on paper or microfilm makes a great difference. An 8½×11-inch page can be transmitted to the user's desk on a LAN in two or three seconds. With paper and facsimile systems, it takes 15 to 20 seconds.
- *Workflow (prioritizing, tracking, and reporting).* Many business functions and activities are paced and sequenced by the flow of paperwork, not by the actual work done with the documents. For example, processing tasks are started when the appropriate paperwork arrives on an individual's desk, not when the previous process is complete. By using electronic imaging systems to manage the paperwork flow and control the pacing and sequencing of paperwork, individuals using the system can be automatically prompted to begin the next procedure. Time-sensitive documents can be expedited through the system by being assigned a high priority. An electronic audit trail and management reports can be established to review the progress of work.
- *Indexing and cross-referencing.* Electronic imaging systems provide the ability to do cross-indexing of documents by computer. For example, users can search for the document by name, account number, key words, or any combination of these. This function can be done with a computer for paper, microfilm, or magnetic media as well, but in electronic imaging systems this function is integrated with other processes.
- *Integrating with other systems.* Because electronic imaging systems are digital systems, they can interface with other digital systems, such as

data processing systems and facsimile systems. A user can retrieve documents relating to an account, for instance, from the optical disk while simultaneously accessing related data from the company's mainframe. The same monitor can display both images and data using windowing techniques.

- *Good image quality.* High-resolution monitors are readily available. Laser printers provide good contrast, high-resolution copies on plain paper.
- *Document always available (displayed on more than one monitor).* It is very time-consuming and frustrating to search for a document that is misfiled. There is also the problem of more than one person needing the same file or document at the same time. With optical disk systems, many people can use the same file or even the same document simultaneously, and documents and files cannot be misfiled.
- *Elimination of head crashes.* With a system that uses magnetic media, a "head crash" of the hard disk can destroy all the data and be devastating. The laser heads of an optical disk in an EIM, on the other hand, rely on a laser beam that can be located further from the disk and therefore there is no risk of a head crash.
- *Security.* Access can be restricted by establishing a user password system for various categories of material. If required, an audit trail can be maintained relating to the access to a particular file or document. Duplicates of the media (optical disks) and the index can be stored at a remote location for protection against disasters.
- *WORM disks cannot be changed.* Once a document is converted to digital form and written into a *WORM* (write once/read many) optical disk, it cannot be changed. The fact that a document cannot be altered is an advantage in many records management applications. Many other types of optical disks are more difficult to alter than paper or microfilm. (See chapter 3 for more details.)

The reason EIM systems are so attractive is not any one of these advantages, but the sum of them. Herb Schantz has come up with a series of reasons why senior management will endorse an electronic imaging system. His 1991 AIIM Resource Report, "Optical Digital Imaging Text Systems" enumerates them. Schantz points out that an EIM system lowers the business breakeven point by reducing operational costs. Much of this reduction is in labor costs. For example, mail handling and document preparation, batch and data preparation, data capture, document storage and retrieval, data processing and customer service, data and image auditing and quality assurance would cost $243,100 under a manual system, but $175,900 under an electronic imaging system doing the same work. Schantz concludes that an EIM will be cost-justified within a 36-month payback period. An EIM with OCR would reduce costs for equivalent work to $121,900.

EIM, Office Management, and the New Office

Only the most short-sighted manager would see EIM as a totally independent phenomenon. Instead, managers should view EIM in the context of the progressive ideas forming a philosophy called "the new office management." Twenty-five years ago, office management involved supervising bookkeepers, file clerks, data processing operators, stenographers, and other clerical staff. Since then, the term has evolved to become "information management." Information management describes the administrative functions necessary to manage information. The emphasis is on controlling information flow through its lifecycle, rather than on controlling people. The basic stages in an information lifecycle include the following:

- Information reception or creation
- Processing and use in decision-making
- Eventual elimination or placement in permanent archives

The excitement over the realities of "information management" seems to be contagious. Offices, when planned and managed well, can be efficient administrative centers. These administrative centers can help private-sector organizations maximize profits. They can also help public-sector organizations get maximum use from their limited budgets. And yes, morale can run high in these environments so that personnel find their workplaces pleasant, satisfying, and motivating.

The change in rhetoric is also accompanying a change in the physical and psychological realities of offices themselves. Traditional "cubicle" offices with receptionists and secretaries guarding executives are becoming less common. Eventually, they might become extinct. Meanwhile, the traditional "line" organization, with authority flowing from top management down to production and administrative personnel, is starting to fade.

The new office involves the emergence of both new technologies and new philosophies. Micrographics and computerized management are working toward the goal of replacing hardcopy paper records with more efficient automated means. Similarly, new managerial philosophies seek to replace old offices that evolved into their present state usually without the benefit of scientific planning, organization, or control.

That is not to say that you cannot have a traditional office using EIM techniques. Even today, many managers, administrators, and other office personnel continue to work in relatively traditional organizations using EIM methods. You do not have to adopt a New Wave philosophy to use EIM equipment and other technologies. Experts agree that we have already reached the point where everyone working in an office needs to have a broad understanding of information, considered one of our most-needed resources. Information management can be an empowering force. The elimination of paper shuffling frees up more people to become involved in strategic planning, organizational, and human development issues.

Inevitably, process design changes the work environment. There is no way of denying the simple fact that document image processing eliminates or diminishes jobs related to the search and retrieval of paper. Edwin McDonell, partner in the Indianapolis-based management consulting firm of Crowe, Chizek, and Company, compares these works to binders who once stood around tables, handstitching books together. The introduction of collating and binding machines eliminated these jobs.

"In imaging, we are looking at whole groups of positions of paper shufflers who will disappear," McDonell states in "When Paper-Handling Jobs Go Away" (*Inform*, February 1992). The good news, he adds, is that "they will be succeeded by individuals who are not only handling paper, but making decisions."

The Bank of Boston, for example, cross-trained its staff of paper handlers and data entry operators to become imaging-literate terminal operators. Staff sizes, thanks to promotion and attrition, whittled down from 41 to 24 people. The survivors had new skills as well as entirely new career tracks. Similarly, the Office of Vital Records of Health and Human Services in Houston, Texas was able to upgrade existing positions when it eliminated two shelving jobs after converting to an imaging system. Computer literacy is now a requirement for career advancement there.

Arthur Falvain, a system consultant at the Bank of Boston, explains the ramifications of the changes: "Some people were considered experts at finding things, a skill not needed any more. These positions are still looked at as clerical jobs, but the actual work they're doing is nearly expert work. It's a very professional operation now, because people can spend their time talking about things that really matter—customer service, resolving problems—rather than spending their time looking for things."

Companies can receive immediate financial rewards because of this *paradigm shift* toward more managerial employees thanks to EIM. Northwest Airlines, for example, now has a higher percentage of higher job classifications in its unionized auditing group. Before the change, auditors would project revenues from a small sampling on flights. Thanks to added personnel punch, however, the group can now audit all ticket sales. Fare violations are down. The company now has accurate market information to help set the best possible seat prices and schedules for various markets.

EIM has also created a new classification of paper-handling jobs: scanners and indexers. Scanners, of course, input documents. The task requires considerable expertise. The Florida Department of Education, Bureau of Teacher Certification, recognizes this fact by offering scanner operators the best paraprofessional grade in state government. In many cases, the whole job classification is new to an organization. Indexing, on the other hand, requires steps such as:

- Identifying the type of document
- Identifying a document's origin
- Matching the customer number or name
- Identifying the electronic file to which it will be linked

In many cases, indexing documents can use the same technologies as mail handling. The same basic procedure for writing a code on the bottom of a letter can be used to write a document on a computer system.

No one, however, regardless of how strong their advocacy for EIM technology or new office management strategies, can ignore the fact that innovation requires adjustments. Consultants such as Helene Roos, in San Francisco, California, focus on managing the human side of technological change. Such consultants emphasize the importance of cooperation within an organization to ensure the adoption of a new system.

"Business process redesign is not enough," Roos says. "People must be prepared for the change process and the organization's culture must be built or modified to support imaging." Roos explains that in order for a three-department project to work within a state government, she had to refocus all three departments away from their interagency conflicts and "fly a banner of tri-agency success."

Chapter

3

Optical Storage

The optical disk evolved from the video disk, which was first introduced for prerecorded motion used for entertainment and training and later for music. All optical disk systems use binary digital coding to store the information, a high-powered laser beam to record the information, and low-powered laser beam to read back the information.

Four types of optical disks are used for documents:

- CD-ROM
- Write-once
- Rewritable
- Multifunction

These media tend to be more costly than magnetic disks and have slower access speeds. At one time, analysts predicted that they would replace magnetic disks, but this has not been the case. In fact, a 1993 Datapro survey found that three forms of optical disks were used by more than 70 percent of all vendors, but 77 percent of these same vendors used magnetic disks and 37 percent used magnetic tape. In fact, image index databases are often stored on magnetic disks separate from the optical disk containing the images.

CD-ROM

The typical CD-ROM (compact disk-read only memory) is 12 centimeters (4.72 inches) in diameter. The primary application for CD-ROMs is publishing. Typical publications are catalogs, directories, encyclopedias, and other reference materials. A publisher replicates thousands of disks and distributes them to users much in the same way reference works on paper are disseminated to users. The user may read this type of disk thousands of times with the appropriate hardware and software. There is no wear to the disk regardless of how many times it is read, since it is read with a laser light source. The information on a CD-ROM cannot be changed.

CD-ROMs are sold by many publishers with thousands of titles being available, including Grolier Encyclopedia, Moody's Investor Service, Oxford English Dictionary, and BiblioMed Citation Series. These publications are available to the public at a price.

CD-ROMs are also used for proprietary reference material. The CD-ROMs in this case may be produced commercially or by the user's own organization. Mack Truck disseminates a CD-ROM parts catalog to its repair stations worldwide. Arthur Andersen publishes accounting reference material on CD-ROMs for its auditors. These publications are proprietary and are only available on an internal company basis, not to the public.

Multimedia presentations also lend themselves to CD-ROMs. For example, the *compact disk interactive* (*CD-I*) standard covers data, audio, and video. Currently, however, it is not widely used. The *CD-XA* (*compact disk-read only memory, extended architecture*) adds color images and several levels of sound quality. IBM has endorsed this standard. Moreover, the Multimedia Product Council has declared CD-ROM the distribution medium for the multimedia personal computer (*MPC*).

The production of a CD-ROM takes place in three stages: premastering, mastering, and duplication. Today you can premaster and master a CD-ROM for less than $3,000 in less than five days. Some systems, such as Meridian Data's VR Pro, can produce special writable disks in minutes. The $50 to $80 disks created by these mastering systems are best used for small disk runs or for checking masters.

The CD-ROM technology patent is owned by Sony and Philips and licensed to other manufacturers. For example, Kodak uses Philips drives for its Photo CD. These drives, by the way, can be connected to a standard television set. The inexpensive cost of a Kodak Photo CD-ROM (around $5.00 when introduced) promises to help bring down the price of other writable CD-ROMs.

Cost is one of the key reasons for the rise of CDs. Making duplicate copies in other media is expensive. Copying CDs, on the other hand, is an inexpensive process. At the time of this writing, small-run CD-ROMs were selling for $1.60 each, and most analysts expect this price to go down. This low cost is why the technology is emerging as an attractive way of disseminating information to a wide audience. In fact, CD-ROM was the first of the nonmagnetic data storage technologies to reach the marketplace. It has since gone through several quantum leaps of improvement. Years ago, recording data on CDs involved considerably more error correction, larger data blocks, and slower access and transfer rates than magnetic storage devices. This is no longer the case. As a result, prices have dropped significantly, and sales have gone up.

CD-ROMs require special players capable of reading them. A CD-ROM player reads data from prerecorded compact disks using a low-powered laser that detects variations in the premastered medium. If you examined a CD-ROM under a powerful magnifier, you would see spiraling reflective tracks. There are approximately 16,000 tracks per inch of surface. These tracks, in turn, are interrupted by pits. The laser reflects light off the reflective surface to a light detector. When the beam crosses a pit in the track, it is scattered. A CD-ROM player reads data from prerecorded compact disks. It contains a special digital-to-analog converter and includes error detection and connection circuitry. The number of applications using CDs continues to rise, albeit slower than industry players and market analysts project.

CD-ROMs do have their drawbacks. Drives are relatively slow. A typical CD-ROM drive will take 300 ms to retrieve data—a speed that is approximately 20 times slower than high-capacity magnetic drives. A more important source of delay is the wait for updates while masters are created and distributed to users. Individual CD-ROM duplicate disks cannot be updated. (Some users prefer online information services for updating records because they provide instant access.)

An individual 4.7-inch CD-ROM will hold 635M of data. About 82M are used for a variety of computer functions, leaving 553 usable megabytes. Approximately 77 of the 82M used for computer functions goes towards allowing the drive to recover from small interruptions in the datastream. The error rate for reading CD-ROM ranges from 1 in 10 billion to 1 in 10 trillion, or approximately one error per 18 disks. You can also find portable 3.5-inch CD-ROMs that hold 180M.

WORM

WORM disks (write-once, read-many) are the most common optical disk technology available today. WORM optical drives function in much the same way as their magnetic counterparts. A magnetic disk works by having a platterlike surface rotated beneath magnetic heads that write and read data on the disk. Optical drives, on the other hand, use carefully controlled laser emissions to write and read data. The biggest advantage of WORM optical storage technology over magnetic media is that it offers a significantly lower cost per megabyte.

In the beginning, however, many potential users were put off by the write-once capacity. You cannot update or add to a WORM disk. However, over time, the computer market has seen more value in having a tamperproof storage medium. In an effort to get the best of both worlds, many subsystems combine conventional magnetic disks with WORM drives.

TABLE 3.1 Disk Types and Sizes, English and Metric.

	Inches	Millimeters
CD-ROM/Audio	4¾ (4.7)	120
WORM	5¼ (5.1)	130
WORM	12 (11.8)	300
WORM	14 (14.0)	356
Rewritable	3½ (3.4)	86
Rewritable	5¼ (5.1)	130

WORM disks are available in several sizes: 5¼, 12, and 14 inches in diameter (see Table 3.1 for metric dimensions). The disks are blank when purchased and are housed in a plastic cartridge. This type of disk is created (written on) by the user with the appropriate hardware and software.

The primary application of WORM disks is all types of business documents, including correspondence, invoices, vouchers, completed forms, and engineering drawings. Like the CD-ROM, a WORM disk can be read thousands of times with no degradation and the information cannot be changed. The main difference between WORM and CD-ROM is the WORM disk is created by the user and the CD-ROM by a publisher. The WORM and CD-ROM disks require different drives to be read.

A key drawback of using WORMs is that vendors are split between standards. Obviously, disk size is an issue. The ANSI X3B11 standard (and its ISO counterpart) permits the continuous-composite formatting that most vendors support. Also, legality is an issue. Many experts believe that the courts will eventually find printouts from WORM disks even more reliable than printout from magnetic disks. However, presently, the courts are evaluating the admissibility of optical disk printouts.

Rewritable or Erasable Optical Disks

Rewritable optical disks use 3½-inch or 5¼-inch diameter disks in cartridges. A typical application would be frequently changed and updated documents, such as legal documents or engineering drawings requiring many drafts. These disks can be created and changed by the user. The technology used in commercially available rewritable systems is magneto-optics, which uses a magnetic field and a laser light source to change the polarity (positive to negative or negative to positive) of each spot that represents black and white pixels on the document.

The technology is similar to WORM disks, but with one key difference: it provides the write-many-times capability of traditional hard disks. Like WORM drive cartridges, erasable storage consists of magnetic recording material and a protective layer housed in a plastic case. Unlike WORM cartridges, however, the recording layer of some erasable cartridges can be altered when heated by laser to a high temperature. During the recording process, the laser heats the area of the platter to be recorded on. When heated, the recording layer is sensitive to magnetic forces, and the magnetic write head of the drive first erases and then records data. When cooled, the recording layer magnetically stabilizes. During the reading process, the laser focuses on the recorded area and detects changes in polarity of the light reflected from the recording surface.

The Curie point is the phenomenon making this possible. Magnetic materials heated to the Curie point realign their polarity to that of a nearby magnet. The read/write head of the optical drive heats the area being written while exposing the same area to a magnetic field. The surface of the disk becomes more reflective or less reflective depending on the polarity of the field. These patterns of reflective and nonreflective patches are translated into binary codes. A typical magneto-optical drive stores between 256M and 650M per side on a 5.25-inch removable disk, and between 128M and 256M on a 3.5-inch disk. The most common form is the 3.5 inch used by IBM, MOST, and Sony.

Erasable (or rewritable) optical storage offers the same durability, storage, and throughput as WORM technology. Erasable optical technology, while somewhat less mature than WORM technology, is improving constantly. Erasable optical drives offer slightly lower performance than WORMs. They are slower—an average of 25 or 30 ms or more—than high-capacity magnetic disk drives (which can have speeds under 10 ms). Erasable disks also tend to be expensive, although the price of rewritable drives has come down recently. You can now get one for under $2,000, meaning that the price per megabyte is less than magnetic disks. A Ricoh RO-5031E rewritable disk (327M per side) currently sells for $230. A Hitachi OD11-1 (325M per side) currently sells for $180.

Multifunction Drives

A multifunction drive will take both write-once and rewritable disks. As would be expected, they use the same magneto-optical media as most rewritable disks. However, they use an ablative media for the write-once function. Panasonic is noted for creating a one-pass read/write drive which is one-third faster than magneto-optical technology. The Panasonic drive shortens write time by writing directly to the unerased surface. Besides Panasonic, Hitachi, IBM, Sony, and MOST have developed multifunction drives. A Sony SMO-E511 drive sells for $4,895 and individual disks sell for $230.

Optical Disk Capacity

The question of disk capacity is always asked. This, however, is not a simple question to answer. The disk size, document size, compression ratio (which depends largely on the nature of the documents), resolution, and brand all are significant factors in determining the capacity of optical disks. Table 3.2

TABLE 3.2 WORM Disk Capacity.

Disk size	Capacity in MB	Capacity in GB	Capacity in pages 10:1 Compression	Capacity in pages 15:1 Ratio
5¼"	940	0.94	20,000	30,000
12"	9,000	9.00	200,000	300,000
14"	10,200	10.20	220,000	330,000

- 200 dots per inch (pixels)
- 8½×11-inch pages
- Black & white documents
- Two sided disks

TABLE 3.3 Optical Disk Capacity for Engineering Drawings.

	Size	Bits before compression	Bits after compression 10:1	Bits after compression 15:1	Drawings per disk 10:1	Drawings per disk 15:1
A	8.5×11	3,740,000	374,000	249,333	200,000	300,000
B	11×17	7,480,000	748,000	498,667	100,000	150,000
C	17×22	14,960,000	1,496,000	997,333	50,000	75,000
D	22×34	29,290,000	2,929,000	1,952,666	25,000	37,500
E	34×44	59,840,000	5,984,000	3,989,333	12,500	18,750

- 200 dots per inch (pixels)
- Two sided 12 inch disks, 9.0 GB
- CCITT Group 3 or 4 Compression
- Black & white drawings

TABLE 3.4 Engineering Drawing Capacity at 400 dpi.

Size	Bits before compression	Bits after compression 10:1	Bits after compression 15:1	Drawings per disk 10:1	Drawings per disk 15:1
A	14,960,000	1,460,000	997,333	50,000	75,000
E	239,360,000	23,963,000	15,657,000	3,100	4,700

gives capacity figures for all four sizes of WORM disks in bytes and pages. This table assumes general 8½×11-inch office documents, a resolution of 200 dpi and page capacity for both 10:1 and 15:1 compression ratios. It should be noted that pages are page sides, so if a page has information on both sides, it is counted as two pages.

Table 3.3 gives capacity for engineering drawings at 200 dpi, again for 10:1 and 15:1 compression ratios. Since engineering drawings quite often require higher resolution, Table 3.4 gives the figures for A- and E-size drawings based on 400 dpi.

Disk life

Table 3.5 gives the shelf life for media before and after data and images have been recorded. This table provides information for optical disks, magnetic media, and microfilm for comparison purposes. In all cases, it is assumed the media is stored according to national standards and the manufacturers' recommendations, before and after recording. In the case of microfilm, it is assumed that the raw film has a life expectancy of at least 100 years and that it is processed per national standards.

Data reliability

The data error rate on optical disks is less than one in 10^{-12} bits. This means that of the thousands of documents on a single disk containing millions of words and tens of millions of characters and numbers, one character may be incomplete or one line may be broken within the entire disk. In fact, the error rate is almost zero.

Jukeboxes

Optical disks today generally are stored in *jukeboxes*. When an image is requested, the jukebox selects the correct optical disk and robotically mounts it into a disk drive, where a laser reads it. Today, many LANs are beginning to use jukeboxes as backup systems. Optical jukeboxes are a key element of larger systems.

Although in-house integrators can create their own optical jukeboxes, most experts agree that buying integrated jukebox systems is a more sensible approach. If you buy a jukebox, pay particular attention to software. Jukebox software monitors system performance and diagnoses problems. Jukeboxes are subject to more than 20,000 insertions per year, so reliability is also important. Jukebox sizes range from 32G to over 1T (terabyte). They work with both WORMs and rewritable disks. The more advanced jukeboxes have two transport mechanisms (pickers) to reduce backlog time. Prices range from $45,000 for a 261G Cygnet 1602 jukebox to $170,000 for a Perceptics SJB-610.

Optical jukeboxes tend to be slow by computer standards, taking 15 seconds or more to access a disk. Laser Magnetic Storage has developed an alternative to jukeboxes—the LF 4500 Rapid Changer. A Rapid Changer holds 28G, can swap up to 12 optical disk cartridges, can swap up to 12 optical disk units, and can read and write both sides of the disk without removing it and turning it over. You can also combine up to four Rapid Changers to create the equivalent of a single medium-sized jukebox.

TABLE 3.5 Media Shelf Life.

	Mag tape	Microfilm (silver gelatin halide)	Optical disk
Pre-	At Least	2 Years	3–5 Years
Recorded	30 Years		
Post-	At Least	At Least	30 Years
Recorded	30 Years	100 Years	

Digital Paper

Another, less common, optical medium is digital paper (tape). Digital paper, like other optical media, uses a laser beam both to deform and detect deformity of the recording layer. Digital paper, however, deforms a dye-polymer layer rather than a metallic layer like other optical media use. A laser can record faster to digital paper than to conventional WORM media. On the other hand, a lower-powered laser recording to digital paper can actually achieve the same data transfer rates as a higher-powered laser recording using other media. It can also minimize errors.

Most WORM technology relies on sophisticated error correction techniques to minimize the impact of data errors caused by the high recording density. The dye-polymer recording layer offers a lower error rate than a metallic layer, further increasing data throughput.

Future Trends

Many analysts predict the eminent rise of the 12-inch optical disk, which has between 2.18G and 5.6G per side. A 12-disk jukebox for such optical disks could hold 1.4 million pages of documents. The approximate cost of use for these systems is 0.5 cents per page. The capacities of these disks are expected to double in the next two years. Increase beyond that capacity will depend upon the development of 16-nanometer (blue) lasers currently being developed by IBM and Sony. Currently, however, these disks have severe problems. Each drive can only be read from or written to drives from a single manufacturer. Drives are always being updated. However, they do not always read existing materials. Kodak is the only vendor selling 14-inch disks. Each disk contains up to 5.1G per side.

In the near future, you are likely to see a standards squabble between rewritable and multifunction manufacturers. Panasonic prefers its proprietary phase-change technology. On the other hand, LMS and Pioneer favor magneto-optical media for the rewritable function, but ablative media for the write-once function. Hitachi, Hewlett-Packard, Toshiba and others favor the magneto-optical technology and continuous composite format used in most rewritable drives, but with differently coded media for write-once and rewritable disks.

In the long run, new technologies will change the imaging business even further. Since the 1970s computer scientists have envisioned storing documents all at once with laser holography. In a single flash of laser light, the image of a whole page would be stored as a hologram, 1,000 times faster than now—maybe even 10,000 times faster.

Anyone who has ever waited, and waited, and waited for a computer to open a document understands the downside of all-too-common PC bottlenecks. The amount of information the PC can handle is not an issue. It still has to sort and retrieve it one bit at a time. Even for the fastest computers, getting data in and out of stored memory is as slow as the spin of a disk.

Until recently, hologram storage was hampered by a problem with the large crystals used to store the holograms. When data is retrieved, scattered light mixes up information from several pages at a time. However, Lambertus Hesselink, a Stanford professor, is credited with creating a solution that allows laser optics to store more than 50 holograms in a single rod that is smaller than a pin. This rod eliminates "cross talk" between images.

Each hologram, containing between 100,000 and 1 million bits of data, is stored in a rod (also called an *optical fiber*) using two beams of laser light. An *object beam* is bounced off the document or other image to be recorded, before hitting the fiber. A *reference beam*, in turn, crosses the object beam, recording the image in the fiber as an electron pattern. Each reference beam then is encoded so that when bounced into the fiber again, it retrieves only the pattern for the particular page that it stored. In other words, 50 pages of a single document can share the same tiny space, and data can be retrieved from any one with the flash of a laser beam.

Chapter

4

Application Analysis

When considering the acquisition of an electronic imaging system, you must take a hard look at your present document systems—from organizational, user, and economic points of view. You should know your system's present weaknesses, limitations, and requirements before establishing goals and objectives for a new system. The following is a checklist that will provide guidance to any organization making this type of analysis.

Application Analysis Checklist

Organization

- How many people use the information?
- How many user workstations will to be automated?
- What are the goals of automation?
  —Increase professional productivity?
  —Eliminate or reduce clerical staff?
  —Shorten access time to documentation?
  —Improve customer service?
  —Reduce space requirements?
- What are the costs of present methods and systems?

Users

- What needs are being serviced (people and division)?
- What functions should be served within the organization?
- What is the source of the paperwork?

- Is anything currently automated?
- What are today's problems?
- Must your EIM system interrelate with other systems?

Cost justification

To estimate the economic benefits of a potential new system, you must determine the current costs of the various functions. These include the hard, or calculable, costs of items, as well as intangible, or soft, costs.

- Hard Costs
  —People
  —Space
  —Equipment
  —Supplies
- Soft Costs
  —Lost files and documents
  —Low response time
  —Poor monetary control
  —Insufficient information to make decisions

System requirements

What are the shortcomings and limitations of your present operation? What goals and objectives do you have for a new system? The following are seven major areas to consider:

- Customer service
  —Immediate response from document history
  —Shorter processing time for information
  —Accurate history and information exchange
  —Up-to-date information, as and when required
- Competitive advantage
  —Lower cost of doing business
  —Improved efficiency, leading to better customer service
- Telescoping item-processing time
  —Organizational efficiency and discipline
  —Bottlenecks in information flow can be audited
  —Improved productivity through reduced retrieval time
  —Elimination of file-processing delays
- Output document quality
  —More professional product and corporate image resulting from quality output
  —Improved credibility of correspondence through integration of historical information and correspondence

- Collecting quantitative cost-measurement information
  —Transaction-processing time
  —Employee productivity analysis
- Complete document history
  —Up-to-date and always available
  —Track revision history, including annotations
  —Most recent additions immediately online
- Remote access to information base
  —Less mail and express service costs
  —Better document control than that provided by remote duplicate file storage
  —Reduced space, file storage, and handling costs at remote locations
  —Interoffice as well as office and headquarters information exchange

Application analysis takes time, and it takes work. You'll need managerial and user representatives to study your need for an EIM and corporate executives to study the financial aspects. Nonetheless, many of the benefits and advantages of electronic imaging, such as better customer service or a competitive advantage, cannot be measured in dollars.

Backfile Conversion

In almost every application being considered for conversion to an electronic imaging system, there is a *backfile*. The backfile might be a few hundred documents going back a few months, or it might be several million pages of permanent records going back 50 or 60 years. The retention schedule, reference rate, urgency of reference, quantity of material, and operating system should all be considered when determining whether, or how much of, the backfile should be included in the new system. A strategic plan should be developed for handling the backfile in an electronic imaging system.

Should the backfile be converted?

There are at least three alternatives for handling the backfile conversion:

1. Don't convert the backfile at all; continue with paper or microfilm as in the old system. Start the new electronic imaging system from this day forward. There are many applications where the reference rate drops off rapidly within a few months. Both systems would be operated for a short time as the reference to the old systems becomes less and less and the activity on the new systems increases.
2. Convert the entire backfile. This is the only viable alternative in applications where the old records are quite active and this activity continues for many years. In many cases, the application will have a comparatively small ongoing load of documents to scan and the number and speed of

the scanners being purchased is determined by the requirement to convert the backfile in a reasonable period of time. This can be more costly than some alternatives. One alternative to this problem is to rent extra scanners and temporary space, and hire temporary people to do the backfile conversion. Still another approach is to contract with a service bureau.

3. For a large, relatively inactive backfile, scan on demand. Using this methodology, no backfile documents are scanned and indexed into the new system until and unless they are requested. When requested, the document is retrieved from the backfile, scanned and indexed, and made available electronically to the requester. Once scanned, the original document is archived or destroyed, depending upon the organization's policy. This approach often proves the most economical for conversion of large backfile since it will result in not converting a significant portion of the backfile, the documents that are never referenced.

Conversion tasks

There are seven major tasks that must be performed for backfile conversion:

- Document preparation
- Scanning
- Inspection
- Indexing
- Image compression
- Media recording
- Quality assurance

Document preparation or *document prep* takes place before scanning. This function is the manual, tedious job of removing bindings, staples, and paper clips, mending torn pages, etc. This operation, which is quite important, can be very large in a backfile conversion. Chapter 5 provides more information on document prep.

Scanning, inspection, and indexing may be done as one task all by the same person with a document being scanned, then inspected, indexed, and recorded onto an optical disk. Alternatively, a batch of documents may be scanned and inspected with a different person doing the indexing a few hours later (or before scanning from the paper documents). This determination should be based upon the depth of the indexing. If the level of person required to do indexing is much higher than the person to do scanning and inspection, the tasks should be separated. Scanning and indexing are described in more depth in chapter 5.

In a typical system (if there is such a thing), document preparation is done on a batch-mode basis and the documents are delivered to the input workstation. A single page is scanned in one to three seconds. It is then temporarily stored in magnetic memory. Next, it is displayed on the workstation screen and inspected for general legibility and skewing. This takes about one second. If it is not legible, the image is erased and rescanned at a higher resolution (if it was scanned at 200 dpi the first time, resolution is increased to 300 dpi by just pushing a button on the scanner). If the image is acceptable, indexing is done through the workstation keyboard from the image on the screen image (two seconds for 64 characters). The image is now transferred to the optical disk (less than one second). Image compression takes place as part of these functions.

Quality assurance attempts to determine whether the product as produced meet the requirements of its intended use. Quality assurance should not be confused with inspection. In electronic imaging, quality assurance is an audit performed by management. A sampling plan and strategy should be established. Procedures should be developed for sampling and acceptance, and rejection levels set. There are well-established quality assurance techniques that should be consulted in developing a good program. Generally, a number of retrievals are performed in a way similar to that of a typical user. This will test all aspects of the system—hardware, software, indexing, and quality of the images.

Service bureau vs. in-house

As the name implies, a *service bureau* provides service to the user. These services are both tangible and intangible. The tangible services include document preparation, scanning (from paper or microfilm), inspection, indexing, compression, media recording, disk duplication, vault storage of records (paper, microfilm, and optical disks) and shredding of documents.

Tangible services are provided for a fee. Intangible services might or might not be for a fee. Quite often, in order to win a contract, a service bureau will provide certain services at no charge. These services include feasibility studies, systems design, equipment recommendations, training, and consulting.

The other intangible service that bureaus provide is the education of users (customers and potential customers) based on their extensive experience. Service bureaus have done the same or similar jobs for many customers, so the new customer receives the benefit of this experience, which can be significant.

Most service bureaus do work for any firm regardless of industry or brand of optical disk system the user has. Some service bureaus, however, service a niche market, such as law firms. Some bureaus support only certain brands of systems.

Service bureaus claim the following reasons users should consider their services:

- High-speed, high-volume equipment
- Trained and experienced staff
- The space to do the job
- Experience that saves the user learning and research time
- Knowledge of industry standards
- Reduced downtime due to more equipment and larger staff
- The ability to provide a turnkey system
- Elimination or reduction of capital investment
- Predictable costs

Most service bureaus bring the documents to their premises to perform the services. This can be an advantage or a disadvantage. The advantage is that the user does not have to make space available for the operations. The disadvantages are permitting important, perhaps vital, records to go off-site, introducing a delay if a document needs to be retrieved from the service bureau in an emergency. Many bureaus will do the job at the user's site if space is made available—or even in a van in the user's parking lot.

Despite the apparent advantages of service bureaus, some users follow the old adage that if you want something done right, do it yourself. Users claim the following reasons for doing the job internally:

- Knowledge to weed unnecessary material from the collection before scanning
- Greater control over the project
- Knowledge of their organization's overall objectives
- Lower costs (a claim that is most often not true)

If in-house staff are not trained properly or not supervised effectively, quality could be poor. On the other hand, a service bureau must be selected carefully. Electronic imaging will be expensive if ill-managed and unforgiving if quality is compromised. This is true regardless of whether the job is done in-house or by a service bureau.

Sampling, Testing, and Sizing a Project

The magnitude of a document conversion job can only be determined by a large sampling of the documents involved. An extensive sampling of the documents should be made in order to size the job in terms of the cost, staff required, schedule, number and/or speed of the scanners, and number of disks required. This sampling is necessary whether the conversion is done

TABLE 4.1 Document Groups by Quality Level.

Group	Description of documents
I	Books with very good printing quality and high contrast, periodicals and documents typewritten in black.
II	Originals with fine graphics, writing with a lead pencil, and documents with small graphics and high contrast.
III	Ink and pencil drawings, light printing, very fine graphics such as footnotes on the bottom of printed pages.
IV	Manuscripts and drawings with low contrast, graph paper with fine light-colored lines, letters typed with a used ribbon, poorly printed and light or colored documents.
V	Documents with very low contrast.

by a service bureau or in-house. It might be useful to classify the documents sampled by quality level (see Table 4.1).

To obtain a sample, pull one file folder at random from each drawer or each three feet of files. Examine these folders for the following:

- Quantity of 8½×11-inch papers
- Quantity of 8½×14-inch papers (legal size)
- Quantity of pages smaller than 8½×11-inch papers
- Quantity of 11×14-inch computer printouts
- Quantity of bound books and catalogs
- Quantity of oversized folded pages
- Quantity of photos and halftone material
- Quantity of material with information in color, with the color having significance
- The number of pages per document
- Quantity of staples the folder contains
- Quantity of paper clips the folder contains
- Other fasteners
- How many pages are exact duplicates that can be removed
- How many pages require mending
- The general condition of the material: yellowed paper; fragile, old documents; crumpled, dog-eared, stained paper; faded prints

Remember that a page with information front and back should be counted as two pages. The sampling should provide information that can be extrapolated to estimate the size of the job and percentages of the material in each of the categories.

For engineering drawings, note the quantity in each size category. For roll microfilm, microfiche and jackets, how many images are on the average unit? What is the background density and gross fog level of the film? Examine the sharpness of the images (resolution test charts will not be available on some microforms). Squareness of images should also be looked at.

Scanning tests should also be performed. Documents with graphics, documents with lots of small type, and forms take much longer to scan than a simple letter or memo. These tests, using a stopwatch, should be performed with the scanner and compression software that you intend to use for the actual conversion. The scanning resolution should also be the same as planned for the actual system.

Chapter

5

Image Processing

Of all the elements in electronic imaging systems, document preparation is the least talked-about activity, most underestimated in magnitude, and most often does not have enough funds budgeted. EIM should not be cast as a cure-all that will miraculously eliminate all document problems. Document prep is a good example of a task that is labor-intensive and won't go away, even with the most state-of-the-art automated system.

Document Prep

Most document preparation can be handled either by in-house staff or by a service bureau (Figure 5.1), with the possible exception of housecleaning (weeding) a collection, which often requires in-depth, technical knowledge of the company's business.

Weeding involves removing and destroying exact duplicates, unimportant material, and material past the retention standard. Weeding must be done very carefully and by knowledgeable people to avoid destroying what could turn out to be an important and valuable item. In many applications, weeding would be so time-consuming and expensive that it is not done; every scrap of paper in the file is scanned. A few extra images in an optical disk system are not very costly. On the other hand, if the percentage of unnecessary material is high, its scanning will increase the overall capture burden as well as reduce retrieval response time of the system.

Sometimes, a service bureau can do the document weeding, especially when the elements to be removed can be clearly identified, such as the removal of nurses' notes in hospital patient records. It is not unusual for an

Figure 5.1 Employees doing document preparation. *(Courtesy of DAC)*

organization to recognize the need to weed the file before scanning but lack the trained staff to do the job. They must then choose between scanning everything or giving the service bureau very detailed guidelines for removing unnecessary material. It can be helpful to have the service bureau do this work on the users' premises and provide someone to answer questions.

Document prep is important to ensure that all documents required for business or legal purposes are indeed scanned. Since scanners use paper transport mechanisms similar to photocopiers and facsimile machines, document preparation is important for removing staples, paper clips, pins, or other fasteners, which can cause serious and expensive malfunctions. Mending torn documents is sometimes required, particularly when the scanner uses belts or rollers in the transport system. Repairing torn documents is less important with flatbed scanners. Stuck pages must be separated and volumes must be unbound whether you use rotary scanners or flatbed scanners. Oversize sheets, beyond the throat width of a rotary scanner, must be separated and given handling. Although flatbed scanning can better handle oversize sheets, it is much slower then belt or roller mechanisms.

If the paper document contains information whose meaning cannot be communicated in black and white, note this on the paper prior to scanning so the meaning is retained. For example, accounting figures may contain negative amounts represented in red on the original document. Since most scanners will only recognize black and white, written identification of what is red would have to be made on the original record prior to scanning. Some scanners are blind to certain colors. Pay attention to a scanner's *drop-out color* to make sure information is not lost. The drop-out color feature may be useful in not recording certain forms.

In some applications, after the document is scanned, it is copied onto a particular optical disk based on certain factors. Make sure to flag such material before giving it to the scanner operator. For example, a company's vital or permanent records might be put on one disk, documents with 25-year retention on another disk, and documents with less than seven year retention still another. This type of organization makes it easier to destroy appropriate documents according to retention schedules, while protecting vital records.

Digitizing

There are three types of document scanners: vector scanners, which are used in CAD (computer-aided design) systems to create engineering drawings; OCR (optical character recognition) scanners, which convert each number and letter in text to an ASCII code for data and word processing; and raster scanners (digitizers), in which the image is converted to a bitmap. In electronic imaging systems for business documents that use WORM erasable optical disks, raster scanning (bitmapping) is the way documents are converted to electronic binary digital code. OCR techniques might also be used in these systems for indexing the documents.

Raster scanning

Raster scanning, usually referred to as bitmapping, treats an image or document as a rectangular array of points, or pixels. In this process, binary digital coding is used to represent the black and white pixels, with white pixels represented by zeros and black pixels represented by ones. Thus, a black and white document is represented by millions of zeros and ones. It is important to note that a bitmapped file is not meaningful to the computer. The computer knows the document as a large number of zeros and ones which correspond to the document's light and dark spaces. Tonal material, photos, and colored images can also be scanned with a slightly more complex extension of this process.

Image quality

The quality, or *resolution*, of a scanned image is measured in pixels per inch or dots per inch (dpi). When scanning documents, you have some flexibility in determining resolution or number of pixels per inch. Typical systems permit 100, 200, 300, or 400 pixels or dots per inch. The more pixels, the higher the quality (see Figure 5.2).

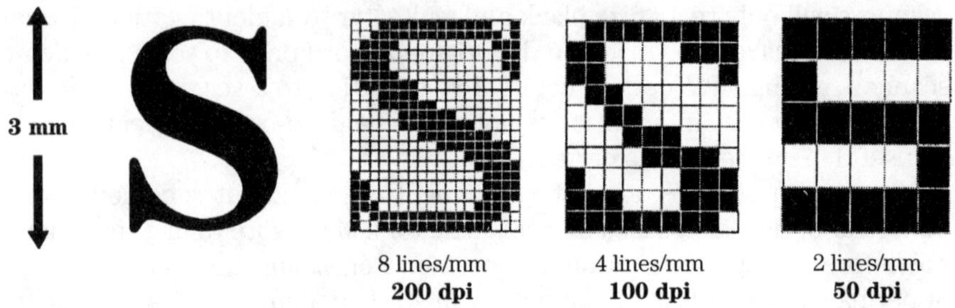

Figure 5.2 Effect of resolution on readability. *(Courtesy of AIIM.)*

There are two trade-offs in deciding what level of quality to select. First, as the image quality and number of pixels is increased, the scanning speed is reduced; if at 200 dpi the scanning speed is likely to be 2.2 seconds per page, at 300 dpi the speed would drop to 3.1 seconds per page. The second, perhaps even more critical, trade-off is the increase in storage requirements (fewer pages per optical disk) as the resolution increases. One 12-inch disk at 200 dpi could store 200,000 pages. If the resolution is increased to 300 dpi, it will only store 100,000 pages, all other factors being equal. The following is offered as a guide to resolution:

Pixels (dpi)	Quality
100	Marginal
200	Satisfactory
300	Good
400	Excellent

More than 80 percent of all documents scanned today are done at 200 dpi.

Scanning preprinted forms

In scanning information on preprinted forms, the variable information typed or written on the form is usually more important than the preprinted information. Depending on the color and density of the inks with which the forms were printed, it might be possible to scan the document so that only the variable data is captured in the bitmapped image. When displaying or printing the document, the variable data can be merged with a stored image of a blank form, to produce an image that contains both the fixed information of the form and the variable information providing a complete document.

The great advantage of this approach is the saving in memory required to store images. Storing a digitized image of both the preprinted form and the variable information that was inserted on the form might require 10 times as much storage capacity as storing only the variable data.

Scanners

From a systems point of view, the electronic image digitizing scanner is the counterpart to a microfilm camera—it does the recording of the information. Most units look and act like photocopiers. There are scanners for different types of input, i.e., document pages, engineering drawings, roll microfilm, microfiche, and aperture cards (see Figure 5.3). A different unit is required for each of these formats.

Scanners are available with multipage automatic feeders, flatbeds for scanning bound material, aperture card feeders, and so on. Most scanners provide controls to change the scanning resolution. Some units can only scan one side of the page, while others scan both sides simultaneously (see Figure 5.4). Scanning speeds for one $8^{1}/_{2}$×11-inch page runs from one to four seconds depending on the machine design and resolution selected. Most scanners have a drop-out color, an image color that will not be recorded.

Compression

Compression is the process of reducing the number of bytes required for storing and transmitting digitized images. Compression uses a technique of counting strings of ones or zeros in a bitmap and, based upon rules in an al-

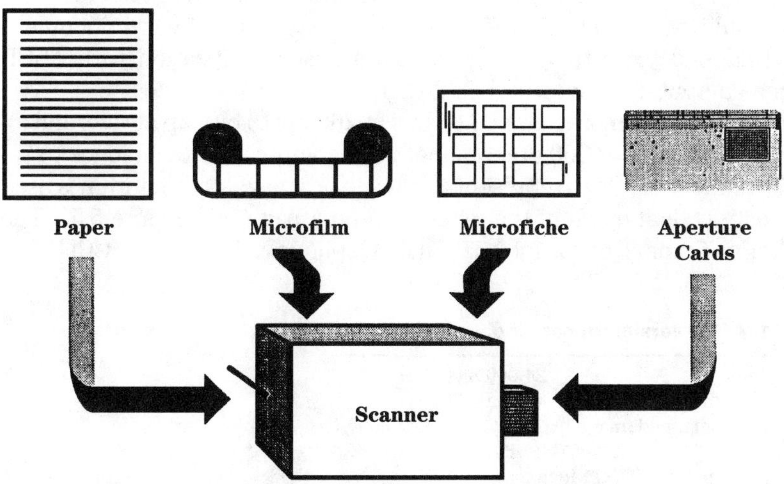

Figure 5.3 Scanners accept many forms of input (not in the same unit).

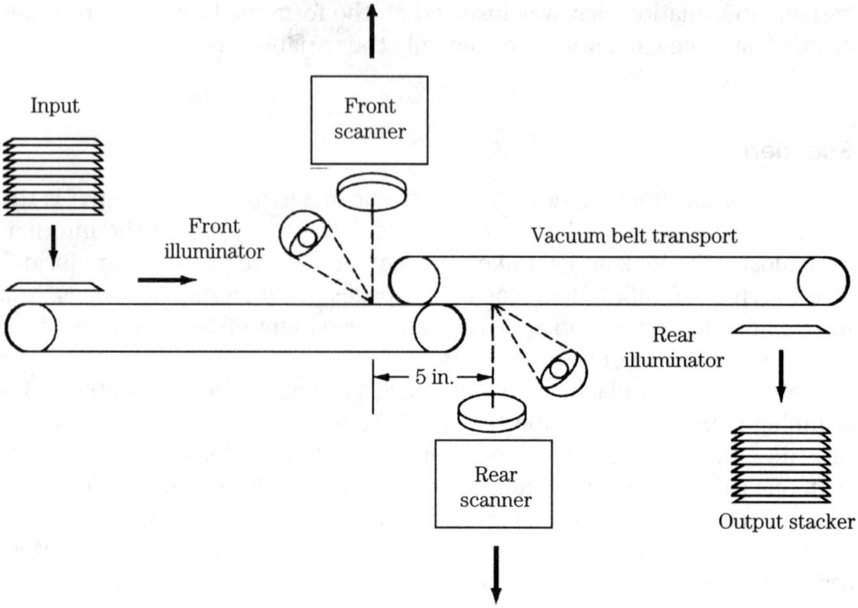

Figure 5.4 Paper transport and imaging system in a typical scanner. *(Courtesy of TDC.)*

gorithm, replacing the strings with coded data to represent the numbers. This technique is very effective on large white or black areas of an image.

An 8½×11-inch document bitmapped at 200 dpi would take 3,740,000 bits, which equals 467,500 bytes (467K). Without compression, it would take this number of bits and bytes regardless of the number of words on the page, or even if there were line illustrations on the page. Without compression, it would take 150 times more memory to store a 400-word document on a bitmapped basis than using character-based digitization with ASCII code in a computer, as shown in Table 5.1.

Because it takes almost four million bits at 200 dpi to bitmap a page, and almost 8.5 million bits at 300 dpi, scanners use compression techniques to reduce the number of bits to represent the image on a page. An algorithm is used to eliminate coding much of the white area on a page (see Figure 5.5). The Consultative Committee for International Telephone & Telegraph (CCITT), a

TABLE 5.1 ASCII versus Bitmapping.

ASCII Code	24,000 bits	= 3 K
Bitmapped at 220 dpi	4,000,000 bits	= 500 K
Compressed bitmapped image	10:1	= 50 K
	20:1	= 25 K
• 400 Word Letter	• 8½×11 inch page	

a United Nations committee, has developed two standard algorithms for facsimile transmission that have been adapted for optical disk systems to compress images. These two standards are designated Group 3 and Group 4.

The exact amount of storage space reduced by compression depends on the amount of text and line work on the document being scanned, the organization of that information, and the compression method used. Compression ratios range from 7:1 to 30:1 in optical disk systems. For example, a simple typed page with a letterhead and signature (see Figure 5.6), scanned at 200 dpi, would take about 132,000 bits when compressed or 3,740,000 bits uncompressed, resulting in a compression ratio of 28:1. A single page in a technical journal arranged in two columns with 9-point type, formulas, and line drawings (see Figure 5.7), scanned at 200 dpi, would take about 230,000 bits when compressed or 3,740,000 bits uncompressed, a compression ratio of 16:1. An $8\frac{1}{2}\times 11$-inch page containing a busy image with more black than white, with Kanji text, as in Figure 5.8, also scanned at 200 dpi, would take about 532,000 bits compressed, which yields a 7:1 compression ratio (see Table 5.2).

Table 5.3 provides a worksheet to calculate the number of $8\frac{1}{2}\times 11$-inch pages in a specific collection that can be stored on a WORM optical disk and the number of disks required for the particular collection. To use this worksheet, determine the quality level and disk size desired, and select and test a sample to determine the average compression ratio for the specific collection.

Table 5.4 is an example of a completed worksheet for 200 dpi, 12-inch disks at a compression ratio of 10:1. Note that if pages other than $8\frac{1}{2}\times 11$ inches are involved, simply determine the number of square inches on the page and replace the 93.5 figure in Table 5.3 with the new calculation.

Indexing

You must establish an accurate index to all the documents stored in an electronic imaging system to facilitate locating, selecting, and retrieving the

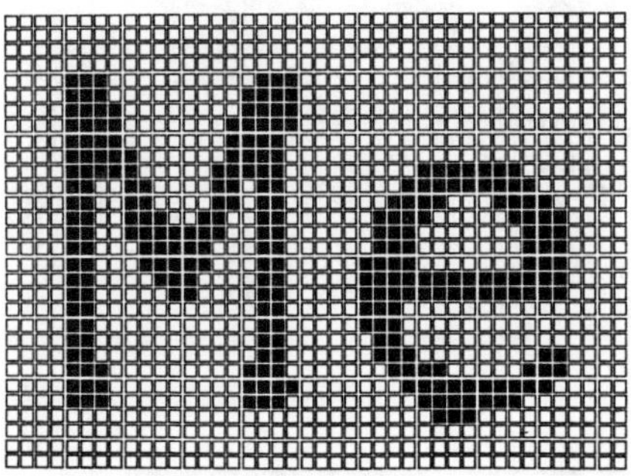

- **Image has 42 × 30 = 1260 pixels**
  1260 bits ÷ 8 = 158 bytes
- **First line 42 white pixels**
  White = 0, 42 zeros
  White run length of 42
- **Second, third and fourth lines = 42W**
- **Fifth line**
  4W, 3B, 10W, 3B, 22W
- **Start of image, 42,EOL,42,EOL,42,EOL, 42,EOL,4,3,10,3,22,EOL...End of Image**

Figure 5.5 A sample compression algorithm.

documents for viewing and copying. Indexing may be very simple or done to great depth. The design of the index is very important, not only because it is the key to finding documents, but also because it has a great impact on conversion schedules and costs.

THE SLEREXE COMPANY LIMITED

SAPORS LANE · BOOLE · DORSET · BH 25 8ER

TELEPHONE BOOLE (945 13) 51617 · TELEX 123456

Our Ref. 350/PJC/EAC 18th January, 1972.

Dr. P.N. Cundall,
Mining Surveys Ltd.,
Holroyd Road,
Reading,
Berks.

Dear Pete,

Permit me to introduce you to the facility of facsimile transmission.

In facsimile a photocell is caused to perform a raster scan over the subject copy. The variations of print density on the document cause the photocell to generate an analogous electrical video signal. This signal is used to modulate a carrier, which is transmitted to a remote destination over a radio or cable communications link.

At the remote terminal, demodulation reconstructs the video signal, which is used to modulate the density of print produced by a printing device. This device is scanning in a raster scan synchronised with that at the transmitting terminal. As a result, a facsimile copy of the subject document is produced.

Probably you have uses for this facility in your organization.

Yours sincerely,

Phil.

P.J. CROSS
Group Leader - Facsimile Research

Figure 5.6 CCITT test page, English letter.

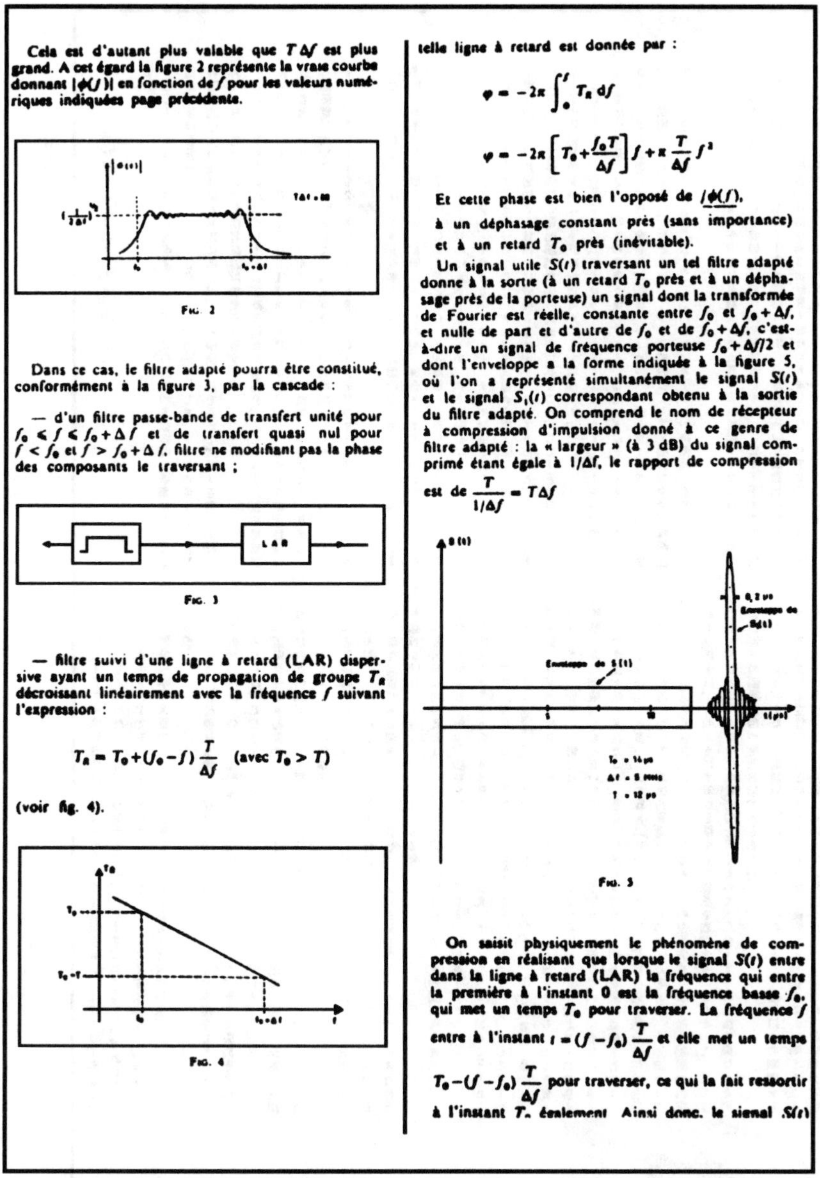

Cela est d'autant plus valable que $T\Delta f$ est plus grand. A cet égard la figure 2 représente la vraie courbe donnant $|\phi(f)|$ en fonction de f pour les valeurs numériques indiquées page précédente.

FIG. 2

Dans ce cas, le filtre adapté pourra être constitué, conformément à la figure 3, par la cascade :

— d'un filtre passe-bande de transfert unité pour $f_0 \leq f \leq f_0+\Delta f$ et de transfert quasi nul pour $f < f_0$ et $f > f_0+\Delta f$, filtre ne modifiant pas la phase des composants le traversant ;

FIG. 3

— filtre suivi d'une ligne à retard (LAR) dispersive ayant un temps de propagation de groupe T_R décroissant linéairement avec la fréquence f suivant l'expression :

$$T_R = T_0 + (f_0 - f)\frac{T}{\Delta f} \quad (\text{avec } T_0 > T)$$

(voir fig. 4).

FIG. 4

telle ligne à retard est donnée par :

$$\varphi = -2\pi \int_0^f T_R \, df$$

$$\varphi = -2\pi\left[T_0 + \frac{f_0 T}{\Delta f}\right] f + \pi \frac{T}{\Delta f} f^2$$

Et cette phase est bien l'opposé de $/\phi(f)$, à un déphasage constant près (sans importance) et à un retard T_0 près (inévitable).

Un signal utile $S(t)$ traversant un tel filtre adapté donne à la sortie (à un retard T_0 près et à un déphasage près de la porteuse) un signal dont la transformée de Fourier est réelle, constante entre f_0 et $f_0+\Delta f$, et nulle de part et d'autre de f_0 et de $f_0+\Delta f$, c'est-à-dire un signal de fréquence porteuse $f_0+\Delta f/2$ et dont l'enveloppe a la forme indiquée à la figure 5, où l'on a représenté simultanément le signal $S(t)$ et le signal $S_1(t)$ correspondant obtenu à la sortie du filtre adapté. On comprend le nom de récepteur à compression d'impulsion donné à ce genre de filtre adapté : la « largeur » (à 3 dB) du signal comprimé étant égale à $1/\Delta f$, le rapport de compression est de $\frac{T}{1/\Delta f} = T\Delta f$

FIG. 5

On saisit physiquement le phénomène de compression en réalisant que lorsque le signal $S(t)$ entre dans la ligne à retard (LAR) la fréquence qui entre la première à l'instant 0 est la fréquence basse f_0, qui met un temps T_0 pour traverser. La fréquence f entre à l'instant $t = (f - f_0)\frac{T}{\Delta f}$ et elle met un temps $T_0 - (f - f_0)\frac{T}{\Delta f}$ pour traverser, ce qui la fait ressortir à l'instant T_0 également. Ainsi donc, le signal $S(t)$

Figure 5.7 CCITT test page, French journal page.

A simple index might be to the folder level. For example, an index of personnel records might direct you to each employee's personnel folder. On the other hand, within that folder there might be an employment application, school transcripts, letters of recommendation, test results, perfor-

CCITTの概要

沿革

CCITTは、国際電気通信連合（ITU）の四つの常設機関（事務総局、国際周波数登録委員会、CCIR、CCITT）の一つとして、ITUの中でも、世界の国際通信上の諸問題を真先に取上げ、その解決方法を見出して行く重要な機関である。日本名は、国際電信電話諮問委員会と称する。

CCITTの前身は、CCIF（国際電話諮問委員会）とCCIT（国際電信諮問委員会）である。CCIFは、1924年にヨーロッパに「国際長距離電話通信諮問委員会」が設置され、これが1925年のパリ電信電話会議のとき、正式に、「国際電話諮問委員会」として万国電信連合の公式機関となったものである。CCITは、同じく1925年の会議のとき、CCIFと併立するものとして設置された。

そして、CCIFは、1956年の12月に第18回総会が開催されたのち、CCITは、同年同月に第8回総会が開催されたのち、併合されて現在のCCITTとなった。このCCITTは、CCIFとCCITが解散した直後、第1回総会を開催し、第2回総会は、1960年にニューデリーで、第3回総会は、1964年、ジュネーブで、第4回総会は、1968年、アルゼンチンで開催された。

CCIFとCCITが合併したのは、有線電気通信の分野、とくに伝送路について電信回線と電話回線とを技術的に分ける意味がなくなってきたこと、各国とも大体において、電信部門と電話部門は同一組織内にあること、CCIFの事務局とCCITの事務局の合併による経費増進等がおもな理由であった。

CCITTは、上述のように、ヨーロッパ内の国々によって、ヨーロッパ内の電信・電話の技術・運用・料金の基準を定め、あるいは統一をはかってきたので、現在でも、その影響を受け、会合参加国はヨーロッパの国が多く、ヨーロッパで生起する問題の研究が多い。たとえば、1960年のCCITT勧告の中で、技術上配慮する距離は約2,500kmであったが、これはヨーロッパ内領域を想定したものである。

しかしながら、1956年9月に敷設された大西洋横断電話ケーブルは、大陸間電話通信の自動化および半自動化への技術的可能性を与え、CCITTがこの問題を取り上げるに及び、CCITTの性格は漸次、汎世界的色彩を実質的に帯びるに至った。この汎世界的性格は第2次世界大戦後目ざましくなったアジア・アフリカ植民地の独立に伴ってITUの構成員の中にこれらの国が加わり、ITUの中に新しい意見が導入されたことにも起因して、技術面、政治面の双方から導入されてきた。CCITTの汎世界化は、1960年の第2回総会がニューデリーで開催されたことにもあらわれている。この総会までは、CCIT、CCIFのいずれにしろ、アメリカやアジアで総会が開催されたことがなく、CCITT委員長も、ニューデリー総会の準備文書で、この点には注目すべきであるとのべている。

任務

ITUは、全権委員会議、主管庁会議を始めとして、七つの機関をもち、それぞれの機関の権限と任務は国際電気通信条約に明記されている。そこで条約を参照してみるならば、CCITTの任務は、つぎのとおりとなっている。

「国際電信電話諮問委員会(CCITT)は、電信および電話に関する技術、運用および料金の問題について研究し、および意見を表明することを任務とする。」（1965年モントルー条約第187号）

「各国際諮問委員会は、その任務の遂行に当たって、新しい国または発展の途上にある国における地域的および国際的分野にわたる電気通信の創設、発達および改善に直接関連のある問題について研究し、および意見を作成するように妥当な注意を払わなければならない。」（同第188号）

「各国際諮問委員会は、また、関係国の要請に基づき、その国内電気通信の問題について研究し、かつ、勧告を行なうことができる。」（同第189号）

上記第187号と第188号にいわれる「意見」とは、フランス語の Avis から訳したもので、英語では、「勧告(Recommendation)」となっている。CCITTの表明する意見は、国際法的には強制力をもたないものであって、この点が、条約、電信規則、電話規則等各国を拘束する力をもっているものと異なる。しかし意見とは称しても、技術的分野では、電信規則のごとき、各国政府が承認してその内容を実施する強制規制をもたないので、実際にある機器の仕様を定める場合には、多くの国の意見が統一されたこの「意見」に従わなければ、円滑な国際通信を行なうことができない場合が多い。この意見（または勧告）は、国際通信を行なう場合当面が直面する問題について、具体的意見を表明するもので、たとえば、大陸間ケーブルで大陸間通話を半自動化しようとする場合、その信号方式や取り扱う通話の種類および料金は、どのようにするかを研究して意見を表明する。したがって、CCITTの活動は、つねに時代の最先端を行くもので、CCITTの活動方向は、そのまま世界の国際通信の活動方向であるともいえる。

この意見は、また、電信規則以下のその他の規則のごとく、数年以上の間隔をもって開催される主管庁会議というような大会議の決定をまたなくても表明することができ、また、その改正も容易であるので、現在のように進歩の早い国際通信界では、関係国の意見を統一した国際的見解としては非常に便利である。

Figure 5.8 CCITT test page, Kanji text.

mance reviews, and medical information. Each document could be indexed. Note that while a document can have many pages, each page is normally not indexed. Some types of documents, such as transcripts of meetings, published articles, and speeches, should be indexed by subjects, topics, or key

TABLE 5.2 CCITT Group III Compression.

	200 dpi 3,740,000 uncompressed bits		300 dpi 8,415,000 uncompressed bits	
	Compressed bits	Compression ratio	Compressed bits	Compression ratio
English Text*	132,034	28:1	225,499	37:1
Journal Text*	229,204	16:1	350,538	24:1
Kanji Text*	531,754	7:1	798,934	10:1

* 8½×22-inch page

TABLE 5.3 Optical Disk Capacity Worksheet.

I - 8½×11 inch page = 93.5 square inches
II - Black text on white background
III - Quality level, *a* dpi (pixels)
IV - *b* inch disk = *c* GB
V - Scanning at *a* dpi = *d* = bits per square inch
VI - One page, no compression, 93.5 × *d* = *e* bits
VII - *e* ÷ 8 = *f* bytes on one page
VIII - *b* inch disk, *c* GB ÷ *f* = *g* pages
IX - *g* × *h* = *i* pages per disk
X - *i* ÷ *j* = *k* number of disks

a - Quality level in dpi (pixels)
b - Disk size
c - Disk capacity in bytes
d - Quality level in dpi squared (times itself).
e - Bits per single page
f - Bytes per single page
g - Uncompressed pages per disks
h - Compression ratio based upon sampling and testing of document collection
i - The number of *your* pages you will be able to store
j - The total number of your pages, divided by the number of pages you can get on a disk, yielding the number of disks you'll need
k - Quantity of empty disks you'll need

TABLE 5.4 Optical Disk Capacity.

I - 8½×11 inch page = 93.5 square inches
II - Black text on white background
III - Satisfactory quality reproduction, 200 dpi (pixels)
IV - 12-inch disk = 9.0 GB
V - Scanning at 200 dpi = 40,000 bits per square inch
VI - One page, no compression, 93.5 × 40,000 = 3,740,000 bits
VII - 3,740,000 bits ÷ 8 = 467,500 (467.5 KB) bytes on one page
VIII - 12-inch disk, 9.0 GB ÷ 467.5 KB = 20,000 pages
IX - 20,000 × compression ratio of 10 = 200,000 pages per 12-inch disk
X - Total number of pages ÷ 200,000 = 12-inch disks needed

words. There are even some documents in which almost every word must be indexed, such as legal depositions or transcripts of a legal proceeding.

The folder-level index will only locate the desired folder; the user will then have to browse the entire folder each time a document in the folder is needed. In this type of system, indexing is minimal and cost of indexing low, but the time of the user to find the desired document is greater than in a system where individual documents are indexed. In designing an indexing system, consider the trade-offs from the point of view of the entire system.

One of the most significant differences among the various electronic imaging systems is the extent to which the information elements contained in the system can be directly accessed. This accessibility relates to the depth of the index. Depending on the nature of the documents and the overall objectives, the significant information elements can be one or any combination of the following:

- Document title
- Document type, such as letter, memo, or voucher
- Author or originating organization
- Recipient or target organization
- Document number
- Date of the document's creation
- Date the document was received
- File classification code
- Retention period
- Destruction date and who authorized
- Key words
- Relationship of the document to other documents or groups of documents

Mechanics of indexing

Document preparation for digital scanning does not require batching as is normally done for paper record systems and microfilming. Generally, indexing is more efficiently done from the image on the display screen than from the paper, although there are many exceptions to this rule. In large backfile conversions, when millions of documents are involved, the most cost-effective indexing method often is from paper. This can actually be done before or after scanning.

Most often in electronic image conversion, the first step is scanning the document and temporarily storing it in magnetic memory. The image, which at this point is not yet registered in the indexing system, is sometimes assigned a control number for registration in an index processing system. In some of the more advanced custom systems, the images are stamped with a control number before scanning. Barcode labels have been used to accomplish this. Barcodes might require special orientation on the document, such as at a 90-degree angle to the motion of the document through the scanner.

Where forms such as purchase orders are used, there might be an identification number preprinted on the document. This ID number, which is always in a predetermined geographical location on the document, can be read using OCR technology from the document bitmap and used for process control or to develop all or part of the index.

Automatic indexing

Basic information about a document, such as customer name and account number, that already exists in a computer database can be exploited in various ways for creating the index or checking index entries. There is no point in retyping this information, since that risks causing new errors as well as adding extra costs and time. Automatic indexing eliminates all or part of the manual indexing process.

Index accuracy

The accuracy of the index is very important in order to retrieve a document from tens of thousands on a single optical disk and perhaps millions in a collection. The fewer the number of elements being entered into the index for each document, the more critical an error becomes. For example, if the application involves purchase orders, and only two elements, *PO Number* and *Date* are entered into the index, an error in the PO Number might mean the document would be lost forever. On the other hand, if the application were insurance claims records with 10 or 12 elements being indexed, and there was an error in the insurance policy number, the document could be found by the policy holder's name or address.

There are several techniques for greatly reducing the incidence of indexing errors, including double-entry keying, the use of check digits on document identifiers, and checking retrieval data against an existing database. Use of automatic indexing and bar coding can help solve this problem. Also, certain features can be built into the system software to reduce errors. For example, if the document identifier should always have eight digits, the system can stop immediately and call the error to the operator if the operator keys in only seven digits.

When should indexing be done?

Document identification and indexing requires knowledge of the specific business documents and more advanced clerical skills. Therefore, in high-volume operations, the steps of document identification and indexing are often separated. On the other hand, the scanning process is a very routine mechanical operation that requires minimal operator skills. There are advantages and disadvantages of indexing before or after scanning.

Indexing before scanning. If indexing requires more than 15 keystrokes for each document, indexing should probably precede scanning. The initial reading for document identification, indexing, and routing is done from the paper copy. Special cases, such as an original form which must be handled in paper (that is, signed and returned), can be identified, possibly with the aid of the indexing programs, and set aside. The documents are then scanned.

An advantage to this is that all special cases relating to document content are handled before the document is scanned. Certain issues of interpretation of the document can be resolved. A disadvantage is that certain manual paper-moving steps which could otherwise be computer-aided are retained through the indexing operation. Another disadvantage is that indexing is separated in time and across computer system boundaries from scanning. Associating the right image with an index entry requires further effort.

Indexing after scanning. An advantage of scanning first is that the quality of the document can be carefully checked, since it must be read from the display as part of the indexing and sorting operation. Moreover, every image in a batch is accounted for through a machine-controlled set of software-implemented safeguards. Images flow from an identity as a batch number to page n of document m, in a natural, quality-controlled way. The switch from manual to electronic processing also occurs at an earlier step, which reduces the paper-handling cost in most applications.

One disadvantage to indexing after scanning is that special steps (batch management) are required in the image system to keep the pages of a document together. Further, the job of defining document boundaries falls upon the indexing personnel. Also, batch management implies that the images in one batch remain together during the indexing process. So indexing must be performed by one person, or the complexity becomes overwhelming. Balancing work among a team of indexers therefore becomes more difficult. Also, if there is a problem or special case, it might be expensive to rectify. Such exceptions tend to be spotted long after scanning, usually when the document is being retrieved by a user.

Chapter

6

Communications

In electronic image management systems, *communication* is the process of electronically connecting the various hardware components of the system. Communication is also the means of delivering documents to users at workstations some distance away from the central imaging system. That might be two or three floors within the same building or thousands of miles.

Analog and Digital Transmission

There are two forms of communication transmission, *analog* and *digital*. An example of analog transmission is a voice telephone call, where the information is transmitted as a continuous waveform. Digital transmission is the transmission of information using distinct on and off electrical states. Recall from chapter 2 that the white areas of an image are represented by zeros and the black areas by ones. Because the computer and electronic imaging hardware works in digital form, and because digital communication is faster and more efficient than analog, almost all communication in electronic image management systems is digital. Although totally digital communication is used in many electronic image management systems, there are many large systems having users and workstations at remote locations that rely on analog telephone systems for communication.

Public telephone transmission systems are largely analog, having been established originally for human voice communication. For communication between electronic imaging components to take place over telephone lines, the digital signal must be converted to an analog signal before it is transmitted. At the receiving end, the analog signal must be converted back to a digital signal so it can be handled by the electronic imaging system. The process of converting a digital signal to an analog signal is called *modulation*. *Demodulation* is the process of reconverting the analog signal back to a digital signal. The device that accomplishes both of these functions is a *modem*, short for modulator-demodulator.

Communication Channels

A *communication channel* is the medium, or pathway, through which information (data and documents) is transmitted between devices. Communication channels fall into three basic types: wire cable, microwave, and fiberoptics.

Wire cable, including telephone lines and coaxial cables, is the most common type of communication channel in use. Because it is easier and cheaper to use the extensive wire-cable networks that already exist, wire cable channels are the most popular. Another reason for their popularity is that the technology used to transmit data over wire cable is standardized, thus reducing compatibility problems.

Microwave signals are transmitted through the atmosphere rather than through wire cables, similar to the way radio and television signals are transmitted. Microwave signals must be transmitted in a straight line; they do not bend around corners or around the curve of the earth. Transmission stations redirect and boost the signals. Satellites are also used to direct microwaves over large, geographically dispersed areas. Compared to wire cable, microwave has a much lower error rate, making it more reliable. Also, because there are no physical links between the sending and receiving systems, communication links can be made over large distances and rough terrains. A disadvantage, however, is the cost of ground stations and satellites to support microwave network.

A fiberoptic system uses light impulses that travel through clear flexible tubing. The tubing is thinner than a human hair; hundreds of them can fit in the same amount of space required for one wire cable. Fiberoptics are very reliable communication channels; information can be transmitted at very high speeds with few or no errors. Unlike wire cables, fiberoptic cables are not subject to electrical interference.

Channel configurations

The two principal communication channel configurations are *point-to-point* and *multipoint*. In a point-to-point channel configuration, a system component (workstation) is connected directly to another device by a dedicated communication channel, giving those devices sole use of that channel. Point-to-point can be an inefficient and costly configuration if a terminal is not active enough to keep the line busy.

An alternative is the multipoint channel configuration, in which three or more devices are connected to the same line. The multipoint configuration uses the communication channel more efficiently and reduces the amount of intercabling needed, thus lowering costs.

What keeps more than one device from sending messages over the communication channel at a time? Two methods are used to determine which device gets access to a channel: *polling* and *contention*. In polling, the controller checks with each device, one at a time, to see if the device has a message to send. If the device has a message ready, transmission begins; if not, the controller moves on to the next device to poll it. After all devices have been individually polled, the process begins again.

The contention method puts the devices in control. Each device monitors the communication channel to see if the channel is free. If it is, the device will send its message. If the channel is being used, the device will wait a predetermined amount of time and try again. This process is repeated until the channel is free. One problem with this approach is that one device can tie up the communication channel for long periods of time.

Channel sharing

Two methods are used to regulate the flow of information from communication channels into a computer: *multiplexing* and *concentration*. Their purpose is to increase the efficiency of a communication channel.

Multiplexing is the process of combining the transmission, bit by bit, from several devices into a single data stream that can be sent over a single communication channel. A *multiplexer* is hardware that produces multiplexing. It is also used at the receiving end to separate the transmissions back to their original order for processing. The rationale for this process is that most communication channels can transmit much more data at any one time than a single device can send. A multiplexer allows the communication channels to be used more efficiently, thus reducing the cost of using the channel.

Frequently, it is necessary to connect more devices than the communication channel can handle at once. Concentration is the process of connecting and serving these devices. A *concentrator*, often a minicomputer, is the hardware that provides concentration. When the number of devices transmitting exceeds the capacity of the communication channel, the information is stored in a *buffer* for later transmission. Many multiplexers also provide concentration.

Common carriers

A common carrier is a company that is licensed and regulated by federal or state government to transmit information that is the property of others at regulated rates. There are thousands of licensed communication common carriers. The largest is AT&T. All common carriers are regulated by the Federal Communications Commission (FCC) and by various state agencies.

Two alternatives to the major common carriers are *specialized common carriers* and *value-added carriers*. Specialized common carriers are companies that supply only a limited number of data-communication services. They are usually limited to communication in and between selected areas. A value-added carrier is a company that leases communication channels from the common carriers and adds extra services to what the common carriers provide. Tymnet and Telenet are both value-added carriers.

Transmission method

The rate that information is transmitted over a communication channel is called the *baud rate*. The baud rate is the number of bits per second (bps) transmitted. Typical transmission speeds are 1,200, 2,400, 4,800, 9,600, and 14,400 bps. Two factors that determine the rate at which information can be transmitted over communication channels are: the bandwidth of the channel and the method of transmission–*asynchronous* or *synchronous*.

Asynchronous transmission is a communication method in which one character is sent at a time. The transfer of information is controlled by start bits and stop bits. Each character is surrounded by bits that signal the beginning and end of the character, enabling the receiving terminal to synchronize itself with the receipt of data on a character-by-character basis. Asynchronous transmission is less expensive than synchronous transmission. It is often used in low-speed transmission on slower speed voice-band channels (less than 1,200 bps), where the transmitting device operates manually or intermittently.

In synchronous transmission, blocks of characters are transmitted in timed sequences. Rather than using start and stop bits around each character, each block of characters is marked with synchronization characters. The receiving device accepts data until it detects a special ending character or a predetermined number of characters, at which time the device knows the message has come to an end. Synchronous transmission is much faster than asynchronous transmission. It commonly uses the faster voice-band and broad-band channels (greater than 1,200 bps) and is usually used when transfer requirements exceed several thousand bits per second. Synchronous transmission is used in electronic image management systems communication because of the high transfer speeds required.

Before information is transmitted, however, a set of traffic rules and procedures, called a *protocol*, must be established. The protocol must be followed by all devices participating in the communication session. The rules of the protocol may vary depending on the devices being used. Prearranged signals defining the protocol to be followed when transmitting and receiving information between system components is called *handshaking*.

Communication channel bandwidths

The bandwidth, or grade, of a communication channel determines the rate at which information can be transmitted over a channel. Voice-band channels can transmit information at rates between 110 and 9,600 bits per second. Telephone lines are voice-band channels. The fastest of these channels is the broad-band channel, which can transmit at rates up to several megabits per second. Advances in technology will soon allow information to be transmitted on some types of broad-band channels in the billions-of-bits-per-second range. Microwaves, coaxial cables, and fiberoptics are broad-band channels.

In all communication systems, there is a direct relationship and trade-off between resolution (dots per inch), speed (baud rate), and bandwidth (the maximum bits that can be transferred along a channel per second).

For example, if the bandwidth of a communications channel is fixed and you want to send 10 pages of documents at a resolution of 200 dpi, it will take a certain fixed amount of time that can be calculated. To send the same 10 pages in less time, you must either increase the bandwidth or decrease the resolution. Conversely, if you want to increase the resolution to get better-quality output, it will take more time to transmit the same 10 pages at the same bandwidth. By increasing the bandwidth, you could increase the resolution or number of documents transmitted at the same time. Increasing bandwidth, however, increases the financial burden.

Networks

A network consists of *nodes* and the links between those nodes. Nodes can be devices such as workstations and printers, or they can be bridges to other networks. The communication between nodes is handled by a combination of hardware and software designed to support a specific set of requirements. Depending on the type of network used and the system requirements, nodes can be connected to each other using twisted-pair cable, coaxial cable, fiberoptic cable, or public telephone lines.

Networks are typically distinguished by the geographic area they serve:

- Local area networks (LANs) serve a limited geographic area, such as an office building or campus.
- Metropolitan area networks (MANs) serve an entire city.
- Wide area networks (WANs) serve a country or the entire world.

LANs

A LAN is a group of computers connected so closely that the machines seem to their operators to be separate doorways to a single bank of data. This quick flow of documents lets a user store or print a file from anywhere on the LAN. International Data Corporation (IDC), a marketing research firm, predicts that the number of PCs connected to LANs will climb to 60 million by 1995. As Mary Ann Richard explained in "Document Imaging Systems and Software: Overview" (*Datapro*, March 1993), LANs are a strong growth area in electronic imaging. She reports, "Most of the new system implementations are PC LAN-based imaging systems involved in better customer service and better use of information resources by professional workers."

One of the key reasons people are turning to LANs is that they maximize low-cost processing power. Indeed, computer magazines are filled with stories about the price/performance trade-offs between LANs and mainframes. No one expects either mainframes or LANs to go away in the near future. However, a LAN server for the IBM SAA Image and Records System is $20,000, as opposed to the $250,000 mainframe version.

Datapro's survey of LAN imaging system vendors found that 91 percent used Ethernet systems. *Ethernet* is a LAN access method allowing connected devices to transmit randomly (broadcast). The problem with this system is that when more than one device attempts to transmit data, their transmissions will collide. Then both devices will wait for different periods before attempting data transmission again. Transmissions, then, tend to degrade performance. On the other hand, Ethernet is fast; it can transmit ten megabits per second. You can eliminate many of the collision problems by segmenting Ethernet into small logical networks. Ethernet is also versatile and uses unshielded phone wire or durable cable.

Token systems are also popular. A full 85 percent of the vendors surveyed by Datapro use *token rings*. Under this system, each device on a LAN can receive a token and send a transmission, or pass the token to the next connected device.

LANs use a variety of operating systems. Novell NetWare dominates the market. Not surprisingly, Datapro found that 71 percent of all LAN vendors surveyed had systems that supported NetWare. The same survey found that Microsoft's LANManager (56 percent) and 3Com's 3COM (45 percent) were also popular. Interestingly, 3Com abandoned its operating system in 1991. Product users were encouraged to migrate to LANManager.

The rise of LANs using products such as NetWare has given birth to the creation of *client/server computing*. At the center of a NetWare system is a single computer known as a *server*. All of the other computers on the LAN are basically system clients. A client/server system ensures that each client has equally easy and complete access to all network resources within the restrictions imposed by network managers. This equality exists regardless of the type of system being used or the location of the server. While each client workstation performs the tasks needed by individual users, processing related to information sharing occurs at the servers. Servers allow LANs to store electronic files in a central location.

Servers come in two types: *dedicated* and *nondedicated*. Most users agree that dedicated servers are the better, safer choice. In fact, some LAN managers strip the keyboards of the PCs being used as servers or unplug them. They plug the computer back in and replace the keyboard only when a server needs service. A nondedicated server can also be employed as a user's workstation. A system with nondedicated file servers requires fewer PCs than a system using dedicated servers, so it tends to be less expensive.

However, nondedicated servers have their drawbacks. Unless the fastest, most powerful, and most expensive PCs are used as nondedicated servers, overall network operations will likely be slowed down more than if they were used as dedicated servers. Also, if someone using a nondedicated server as a workstation has a glitch with an application, that problem could disrupt everyone else on the network connected to that server. If the server-workstation must be restarted, every network operation underway at that time might need to be restarted as well.

Although most office networks use a single server, some have multiple servers. A second server, for example, can store backup copies of all data on the primary server and be activated instantly should the primary server fail. Multiple servers allow network users to continue working while the first server is being repaired and reactivated. These multiple servers can all be centralized, located in the same place for ease of management and repair, or they can be distributed to various locations on the network.

As you probably know, the two most popular types of microcomputers in today's offices are IBM compatibles and Macintosh computers. In the past, the entire computer industry was divided into two camps: IBM supporters and Mac supporters. The IBM PC family, which originated in 1981, is the corporate standard throughout the United States. However, IBM does not manufacture all members of this family. They are manufactured by companies who design their systems to be IBM compatible.

The second most popular family of desktop computers is the Macintosh family, produced by Apple Computer. When it was introduced in 1984, the original Macintosh computer was laughed at by corporate data processing departments, who looked at these user-friendly machines as toys and not fit for serious business applications. However, the system, with its picture-oriented software and emphasis on simplicity, attracted too many converts for corporate computing managers to ignore. Soon, developers of leading business software from the IBM-compatible world began supplying Macintosh versions of their products. Today's crop of Macintosh computers includes sufficient power and memory to run applications not available for some IBM-compatible PCs. In addition, Macintosh computers can now be equipped with disk drives that can work with disks from newer IBM-compatible PCs.

A few years ago, vendors seemed to emphasize a software migration towards standards-based platforms. Today's imaging industry, however, seeks to support mixed vendor environments. Users no longer are married to either the Mac or the IBM worlds, or even the world of UNIX workstations. Instead, they want inexpensive programs that can be accessed by any client workstation in the organization, including Macs, UNIX workstations, and IBMs. For example, Unisys' InfoImage Folder, an electronic folder management system that accommodates large-format and business-sized documents, employs both MS-DOS/Windows clients and UNIX servers.

Fortunately, computer connectivity is easier than ever. To form a LAN, a desktop system and its components all must have physical access to some type of network pipeline. Most IBM-compatible PCs use network interface adapters. These adapters go by such names as network interface cards (or NICs), LAN cards, or network cards.

Network interface adapters cost from $100 on up. The price depends on the adapter's features and manufacturer. It will also be influenced by where you buy it—a retailer, discounter, or mail order operation. In recent years, companies such as Novell and 3Com began lowering prices on network interface adapters.

The network interface adapter usually also contains processors that translate signals as they pass between the network and the connected device. The specifics of the network interface adapter depends on the type of network and desktop computer to be connected.

Network interface cards, by the way, share features with modems. Recall that a modem lets a PC talk on the telephone to another PC, a larger minicomputer or mainframe, or a computing network. Many people use PCs and modems to access networks offering subscriptions, such as GEnie, CompuServe, or MCI Mail. These services provide a variety of business and social opportunities, including course instruction, electronic mail, shopping, and interactive games.

A PC modem can either be internal or external. An internal modem consists of a board of chips and circuits that plugs into an expansion slot inside that PC in the same way that a network interface card does. An external modem consists of a separate small box that plugs into a serial communications port on the PC. Either way, the modem performs the same tasks via a telephone line that a network interface adapter does in a LAN, giving the PC user access to the power of a computing network.

Macintosh computers, unlike most IBM-compatible PCs, come already equipped to participate in networks based on AppleTalk, Apple Computer's own network. Instead of extra network cards, most Macintosh computers are connected to a network via their printer ports, which double as what Apple calls LocalTalk connectors. Unfortunately, this Apple-designed network is slow and somewhat limited. Fortunately, Macintosh computers can be equipped with a network interface adapter like an IBM-compatible PC. Adapters that plug into the Macintosh disk drive port also are available, as are devices that act as bridges from Macintosh-based networks and networks designed primarily for IBM compatibles.

PC LAN imaging systems work best where document handling requirements are between 500 and 2,000 pages daily. LAN imaging systems are not adequate for applications requiring the highest levels of customization and integration. LAN imaging systems can, however, begin as standalones for application prototyping. IBM ImagePlus/2 and Keyfile Corporations' Keyfile are examples of imaging products that cater to this segment of the imaging market.

Network topologies

In a network, every node must be able to communicate with every other node; however, this does not mean that point-to-point links are required. Nodes can communicate with each other via intermediate nodes. The use of intermediate nodes to route messages reduces the number of physical links required and ensures that every communication line is being used as efficiently as possible.

The communication lines and the nodes they link form a geometric pattern known as a *topology*. The topology describes the way in which the nodes are linked to each other. LAN topologies are designed to optimize communication given limited distances, high-speed requirements, and wiring considerations. The most commonly used topologies are described briefly here; each approach has advantages and limitations that govern its applicability in a network:

- *Bus* topology consists of one communication line shared by all nodes. Devices are attached to the line by transceivers, which transmit and receive messages. A source node sends a message that includes the address of the destination node. All the nodes on the line receive the message, but only the destination node accepts it (see Figure 6.1).
- *Ring* topology consists of a series of nodes connected by point-to-point links that form a ring. Messages move through the ring in one direction. Each node receives the message; however, only the node to which the message is addressed accepts it. This topology provides high throughput for networks on which traffic is heavy and uniform (see Figure 6.2).
- *Star* topology consists of a central node to which all other nodes are connected through point-to-point links. The central node controls all message transmissions. This topology lends itself to networks in which there is heavy traffic between the central node and the peripheral nodes, but very low traffic between two peripheral nodes (see Figure 6.3).

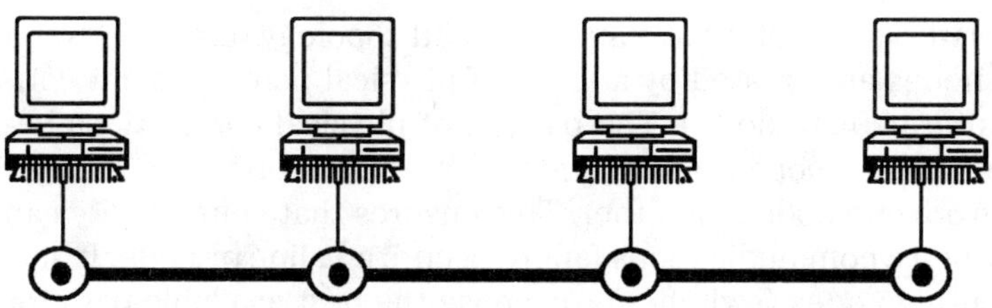

Figure 6.1 LAN bus topology. *(Courtesy of Unisys.)*

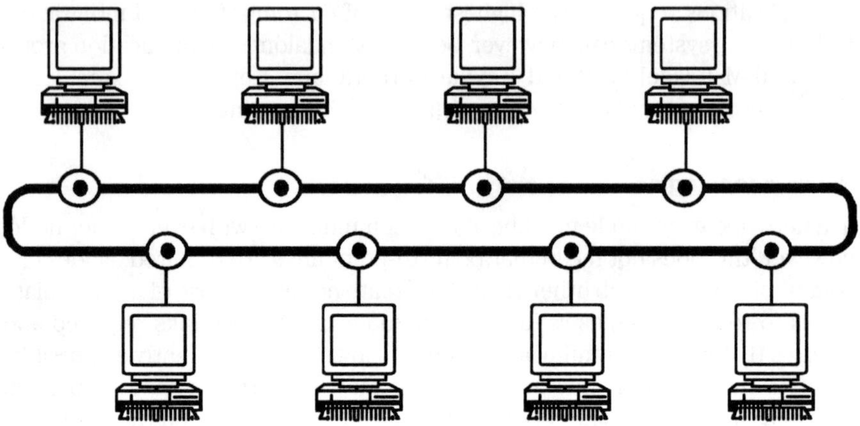

Figure 6.2 LAN ring topology. *(Courtesy of Unisys.)*

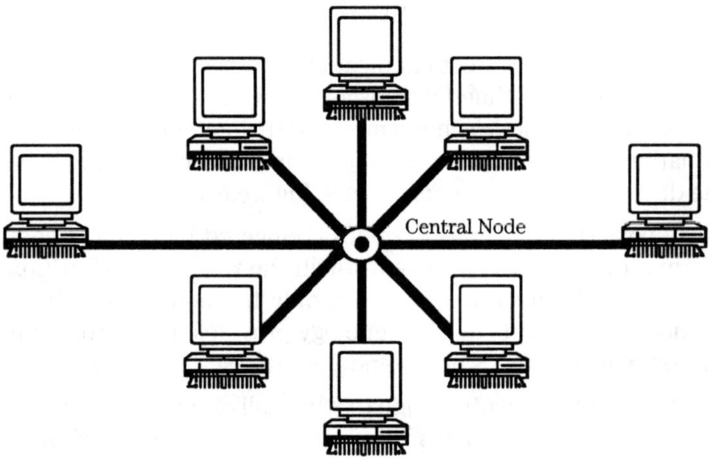

Figure 6.3 LAN star topology. *(Courtesy of Unisys.)*

WANs

Wide area networks (WANs) have an undefined topology. Links between the two end nodes are created by a chain of physical point-to-point connections of intermediate nodes. The pattern of physical connections between nodes is developed to provide more than one way to get information from one node to another. This ensures that connectivity can be retained when a communications failure occurs at a line or node. It also allows the routing nodes (switches) to choose the best available path for the message.

Networks and imaging systems

Typically, the choice of networking techniques and technologies is based on operating requirements and the system configuration. With imaging systems, a major additional factor is the structure and nature of the documents being transmitted. Because a handwritten or typed document that has been bitmapped requires up to 10 times more memory than the equivalent computer-generated ASCII coded document, it will require a channel that is 10 times faster or it will take much longer to transmit. Therefore, the network should be structured to provide optimal throughput for such large files.

Because imaging systems consists of a variety of devices, it is likely that these devices will be connected to a LAN communicating with similar LANs in different parts of a building or at other buildings in different cities. Additionally, because imaging systems are information management tools that deal with diverse types of information obtained from different sources, they must be able to communicate with a variety of systems that store corporate information, and information in other departments that is to be brought into the imaging system.

Software

At the heart of any computer system is software, the set of instructions used by computers to perform tasks. Software performs the following functions:

- Interfaces between the operator, peripheral components, and the control computer's processor
- Regulates the flow of information among the components and the central processing unit
- Translates commands from the operator into specific hardware instructions
- Allocates hardware resources for specific tasks
- Indexes and cross-references
- Provides routing, document flow, and processing through workflow scripts
- Produces statistics and reports

Most systems need three and sometimes four categories of software:

- The operating system, a turnkey software package that provides housekeeping instructions to the computer, device drivers, and hardware component interface software
- Application software, general-purpose programs that perform such tasks as file data management, word processing, and spreadsheet processing, or special programs for a particular vertical industry or organization

- Utility software, programmer workbenches, or toolkits that allow user organizations to modify the systems software
- Specialized software

The operating system

The first layer of the operating system software provides basic instruction to the control computer's processor in the EIM system. Operating systems such as MS-DOS, OS/2, UNIX, and VMS are used in the products of many vendors. This is called *open systems architecture*. There also are proprietary operating systems which are used by a single vendor. Both types of operating systems can be used with PCs, LANs, mid-range computers, and mainframes.

The operating system software is the foundation onto which the application software is loaded. The most popular operating system in use in PCs today is DOS. DOS allows you to start up your computer and run applications programs. An increasingly popular competitor of DOS is known as OS/2. OS/2 allows for *multitasking*, which means that one computer can do more than one job at a time. Windows is another popular program because it allows PC users to work with a mouse and a screen that resembles a Macintosh configuration. Windows NT is the latest version of Windows; it includes many of the features found in OS/2.

Applications programs that run under DOS, OS/2 and Windows include the following:

- Word processing for use in preparing documents
- Spreadsheets for accounting work
- Graphics for producing art
- Databases for storing large quantities of information

Imaging operating systems need not be limited to DOS and OS/2. Windows NT, many analysts feel, has a strong future as an imaging system. Windows NT runs on Intel Microprocessors, Digital Equipment's Alpha series, and MIP's Computer System RISC processors. UNIX, a mainframe operating software that has migrated to PCs, is also a part of the electronic imaging world. Sunsoft's Solaris 2.0 operating system is also important to imaging. It will run on SPARC microprocessors and PCs.

The second layer of the operating system provides the instructions and interfaces between the peripheral components, such as scanners and jukeboxes, and the control computer's processors. This software is necessary to drive and access these hardware components and to assign locations for the images on the disk (by sector and track). Some components come with operating system software that is integrated with the hardware and cannot be changed by the user, other software is independent of the hardware. There is also some operating system software that can be integrated with user-developed application software.

Application software

The first layer of application software in an EIM system usually provides a file management program to keep track of all the images in the system, as well as image storage and retrieval software and image processing programs. This software can provide simple functions such as cross-referencing, linking name to account number, for example.

Document tracking has changed dramatically over the past few years thanks to improvements in application software. A few years ago, document tracking software did nothing more than record the whereabouts of files as they passed through organizational paperwork paths. Today, however, experts agree that the idea of software only tracking files is gone. While tracking folders is still one of its most important functions, document software has expanded to a broader range of applications.

The second application layer provides software for selected vertical industries, such as legal, insurance, financial, manufacturing, and medical. Although there are many off-the-shelf packages of this type available, most users, even in the same industry, have different needs. Therefore a third layer of application software, which is an expansion of the second layer, provides customized software specifically for the user's organization. This may be proprietary because of the type of index required or the nature of the business.

Utility software

Utility software serves two purposes. The first is to provide a workbench or toolkit, often called an application program interface (*API*) allowing user organizations to modify the systems software. These programs translate high-level data-manipulation activities into simple specific instructions. This software might also provide utilities enabling supervisors to monitor the performance of the system and allowing operators to establish schedules and audit trails for work done on the system.

LAN-based server development tools require the use of APIs that allow any processor in a network to get client service from any imaging server on the network. The most common applications development tool used in imaging are APIs to and from existing applications. Also popular are image utilities, menu-driven programming interfaces, workflow scripting language, 4GL programming languages, and programmable device drivers.

A second layer of utility software is provided for the optical disk file system. These programs provide the scheme for allocation tables, directory lists, jukebox locations, and so on to facilitate overall system performance.

Software management

The central management problem facing many organizations involves trying to sort through the dozens of vendors offering numerous features and options as electronic imaging software evolves. EIM packages currently available include software for some or all of the following:

- Indexing
- Scanning
- Image enhancement
- Image file conversion
- Image visualization
- Editing and manipulation
- Video image capture and processing
- Image compression and decompression
- Printing
- Imaging development toolkits and libraries

You can find a variety of shrink-wrapped imaging systems designed to manage document images. Many of them are general-purpose systems, while others facilitate customization. Many also allow users to manage other document types, including word processing documents and data downloads from mainframes. Such a program usually comes with its own database, image-indexing method, image utilities, image device support, and application interface to I/O services. Popular systems include IBM's ImagePlus, FileNet's Workflow Business System, and MicroDynamics' MD MARS. Document management software today can cost anywhere from several hundred to several thousand dollars. This software can perform the following tasks:

- Track active files, inactive files, historical and vital records
- Log in new files or documents
- Index files or documents
- Print file folder labels
- Print bar codes
- Scan or log documents out and in
- Locate files and documents along the paperwork route
- Track non-paper documents, including PC disks, hard disks, microforms, and optical media

- Perform searches by key word or with a full-text index
- Add new fields or redefine existing ones
- Create wait lists and flags for files or documents
- Generate file activity reports by location, department, and user
- Generate activity reports of delinquent and overdue files
- Offer Help screens
- Include security and password provisions for both standalone and network systems

Image index databases are of particular concern. Many are proprietary, which limits the user's ability to tailor them to existing systems. Others use standard databases such as Oracle and Informix, but add proprietary extensions to prevent other developers from using the database's tools to access information for other applications.

A shrink-wrapped image applications system also often provides a programming language. This language is usually a modified or extended version of the language used by the underlying database. The packages provide these languages to enable users to customize the basic management runtime components of their software.

Hardware Components: A Systems Approach

Although there are an infinite number of electronic image management (EIM) system configurations possible, they generally fall into one of two categories: *semiautomatic systems*, sometimes called electronic file cabinets, and *distributed processing automatic systems*, also known as *enterprise-wide systems*. Table 6.1 shows typical system sizes in terms of users, images, and initial cost for various application sizes.

TABLE 6.1 System Sizing.

	Users	Images	Cost*
Single User	1–2	100–50k	$30k–75k
Work Group	3–10	50k–1m	75k–300k
Departmental	11–50	1m–5m	300k–750k
Business Unit	51–100	5m–10m	750k–2m
Enterprise	100+	10m+	2m–10m+

k = 1,000 m = 1,000,000

Source: Dataquest.
*Initial cost including hardware, software, communications, and installation, but not including backfile conversion.

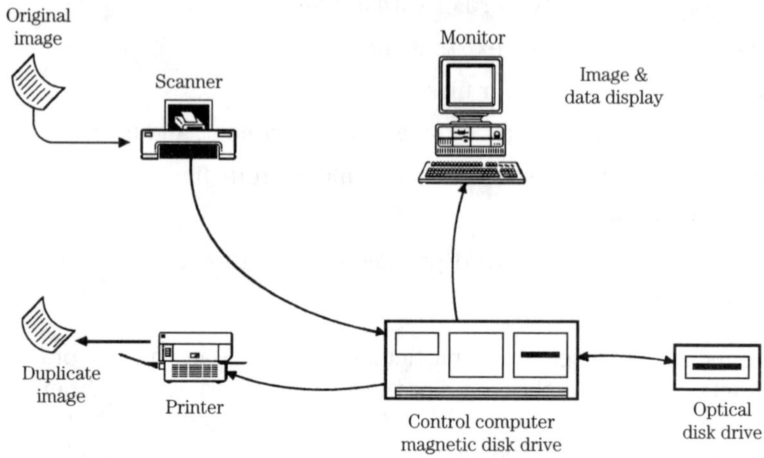

Figure 6.4 A diagram of semiautomatic system (electronic file cabinet).

Semiautomatic systems

A diagram of semiautomatic systems (electronic file cabinets) is shown in Figure 6.4. The following are the functions of each component:

- *Scanning.* The human-readable image (alphanumerics and graphics) on a page is converted to binary digital code in one to three seconds. This process of bitmapping is described in chapter 5. The digitized image is sent to the control computer over dedicated cables.
- *Control computer.* This device acts as the manager of the system, directing the actions of all the components. This computer must have system and application software, as described earlier in this chapter. The computer with the appropriate software provides the interface between the system doing scanning, inspection, and indexing and the users doing retrieval.
- *Monitor.* This device is used for viewing the image for inspection or retrieval. Usually, an entire $8^1/_2 \times 11$-inch page can be viewed. There are also large monitors that can display two pages. The monitor can display information from magnetic memory or from optical disks. In selecting monitors, consider screen size, display mode (horizontal or vertical), resolution, the ability to display multiple images, zooming, rotation, ergonomics, and price (particularly when many are required).
- *Inspection.* The image is displayed on the monitor to check for legibility, skewing, and folded pages. If the image is unsatisfactory, the digitized image (in magnetic memory at this time) is erased and the original is rescanned, perhaps at a higher resolution. If the image is satisfactory, it is usually indexed at this point. The computer, through software, will provide the operator prompts to assist the indexing function. The indexing infor-

mation is stored in magnetic memory; it may also be recorded on an optical disk. The digitized image of the document, which the computer compresses, is transferred to an optical disk. The optical disk is in a disk drive.

- *Disk drive.* A device employing a laser that permits information to be written onto the optical disk (with the laser at high power) or read from the disk (with the laser at lower power). The drive must match the disk size, type, and format. The operator selects the appropriate disk and manually inserts it into the drive.
- *Printer.* A device for making hardcopy of an image, usually on plain paper. In selecting printers, the main considerations are resolution, speed, and price. In the printing process, the computer receives the digital information from the disk drive, decompresses it, and sends the image to the printer. Imaging systems can use laser, inkjet or dot matrix printers. Today, however, the laser printer is considered the standard for business correspondence because of its speed and quality.

It is possible to have a standalone, semiautomatic system with the scanner, control computer, monitor, disk drive, and software all built into a single box. The printer, however, would be a separate unit.

Distributed processing systems

A diagram of a distributed processing automatic system (enterprise-wide system) is shown in Figure 6.5. This system differs from the electronic file

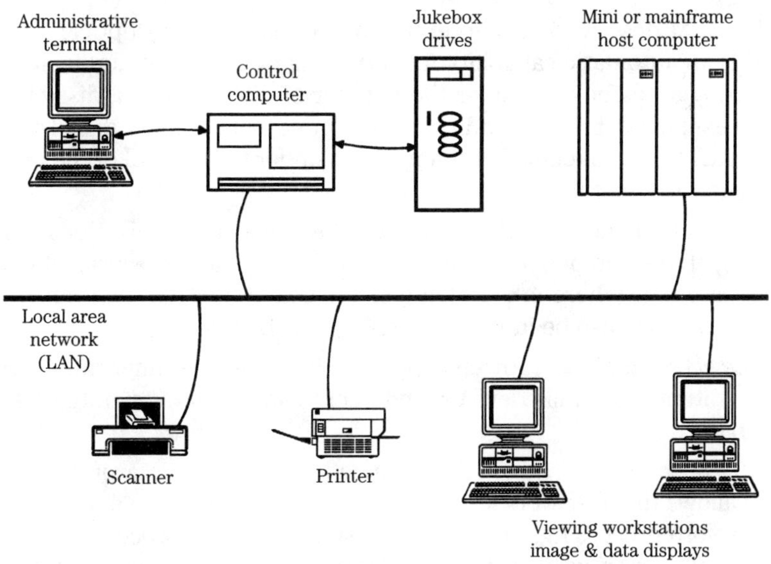

Figure 6.5 A diagram of distributed processing automatic system (enterprise-wide system).

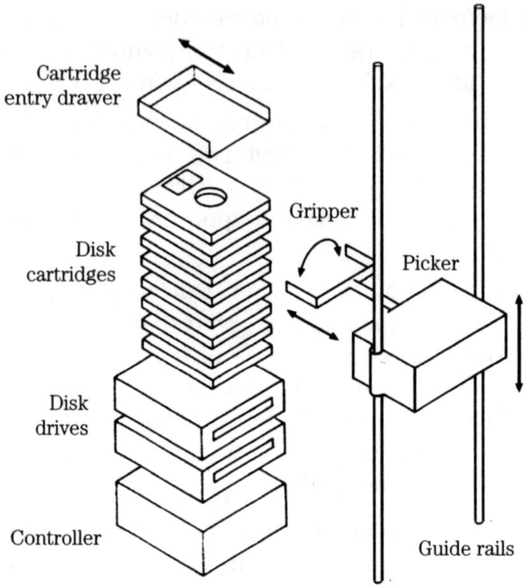

Figure 6.6 A jukebox schematic.

cabinet in that it includes a jukebox, a LAN, many retrieval/viewing workstations, and is connected to a minicomputer or mainframe host.

The following are descriptions of the additional components in an enterprise-wide system:

- *Jukebox.* As described in chapter 3, a jukebox stores many optical disks and adds an automatic capability for retrieving optical disks and images. In selecting a jukebox, consider the number of disks handled, disk size, average exchange time, expandability, and price. One or more disk drives are internal to the jukebox. This unit is also under the control of the control computer (see Figure 6.6).
- *LAN.* As stated earlier in this chapter, a LAN is a communications system tying all the components together and allowing viewing workstations to be located anywhere in a building or on a campus. Multiple scanners and printers may also be interconnected by the LAN.
- *Multiple workstations.* An enterprise-wide system permits hundreds of workstations, facsimile devices, and microcomputers to operate on an online basis.
- *Host computer connection.* The connection to a host data processing system allows the user workstations to retrieve data from the mainframe, perform word processing, electronic mail, and other data processing functions as well as retrieve images all on the same display. Once this integration is made, the entire spectrum of a total information system is possible. The concept of workflow can also be included, as discussed in chapter 7.

Chapter 7

Key Issues

The most basic unit of any computer system is a *standalone*, a machine or device such as an office copier that performs a function by itself. Standalones, however, are only one possible component of a system. A *system* is a set of machines, devices (cameras, processors, readers), and sometimes software in one technology such as micrographics that performs a function. *Integration* is the bringing together of several technologies (such as data processing, micrographics, and facsimile) into a whole, allowing documents and data to flow from one system to another. The integration of systems and technologies forming a total system is made possible through software and communications.

Systems Integration

Any successful organization is the sum of its parts. Each part—an individual, work group, or department—must work in tandem with other parts, including those outside organizations, such as customers and suppliers. A well-run organization has an integrated environment in which resources and information are routinely shared. The resources are held by individuals and groups located both inside and outside of the company. Some are connected by computer systems, others are isolated. Some of the information (data and documents) might be stored in electronic form in online data files. Some might be on magnetic media, some on microfilm. A great deal of it, however, is on paper stored in folders, file cabinets, and off-site facilities.

Systems integration means that the four basic forms of information—data, text, image, and voice—can be exchanged among applications running on a variety of systems and subsystems. From another point of view, integration means combining subsystems—data processing, word processing, electronic mail, electronic imaging, facsimile, and voice annotation—to provide information in any form, instantly, to any user from any subsystem on a single workstation.

The user should not care, nor need to know, how the document was originally created (by hand, typewriter, computer, or whatever). The user should not need to know how the document is stored (on microfilm, magnetic media, or optical disk). The user simply requires the correct information (document or data) on his or her desk when it is needed—that's the job of a good integrated information management system.

Typical Example of Integration

Suppose a toaster manufacturer has an integrated EIM system with correspondence from customers being scanned when it arrives in the mailroom. Today, they received a letter from a customer complaining about the Model 8B toaster that he purchased three months ago for $50. The problem with the toaster is that regardless of the setting, the toast burns. The letter is scanned in the mailroom and electronically routed to a customer service clerk to investigate and prepare a reply. Figure 7.1 shows the letter on the screen of the customer service clerk's workstation.

After reading the letter, the clerk would like to know some basic facts about this model toaster, as follows:

- When was this model toaster manufactured?
- What was the list price of this toaster?
- Have there been other complaints about this model?

This type of information is maintained in the company's mainframe computer. The clerk, using the same workstation displaying the electronic image of the customer's letter, interrogates the data files in the mainframe. The electronic imaging system and the data processing system are connected by software and communications. This particular system has a windowing capability. Figure 7.2 shows the workstation screen with both the image of the letter and the related data. The customer service clerk now can see that the Model 8B toaster has not been manufactured in more than two years and the list price was $40. There has never been a complaint about toast burning on this model toaster. Again, using the same workstation, in a third window, the clerk uses word processing to prepare a reply to the customer. Figure 7.3 shows the workstation screen with the customer's letter, the data from the mainframe, and the reply in a word processing window.

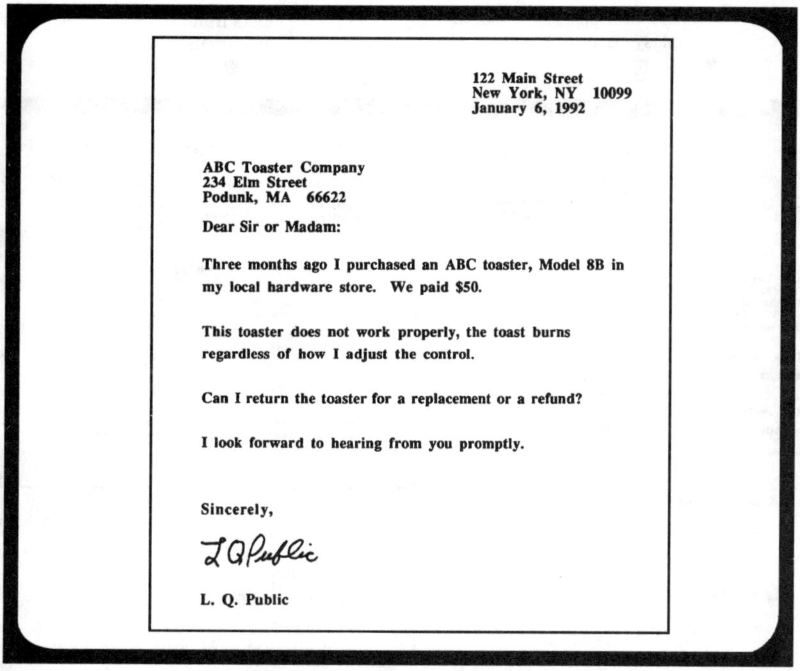

122 Main Street
New York, NY 10099
January 6, 1992

ABC Toaster Company
234 Elm Street
Podunk, MA 66622

Dear Sir or Madam:

Three months ago I purchased an ABC toaster, Model 8B in my local hardware store. We paid $50.

This toaster does not work properly, the toast burns regardless of how I adjust the control.

Can I return the toaster for a replacement or a refund?

I look forward to hearing from you promptly.

Sincerely,

L. Q. Public

Figure 7.1 Image of a customer's letter as it would appear on a clerk's display.

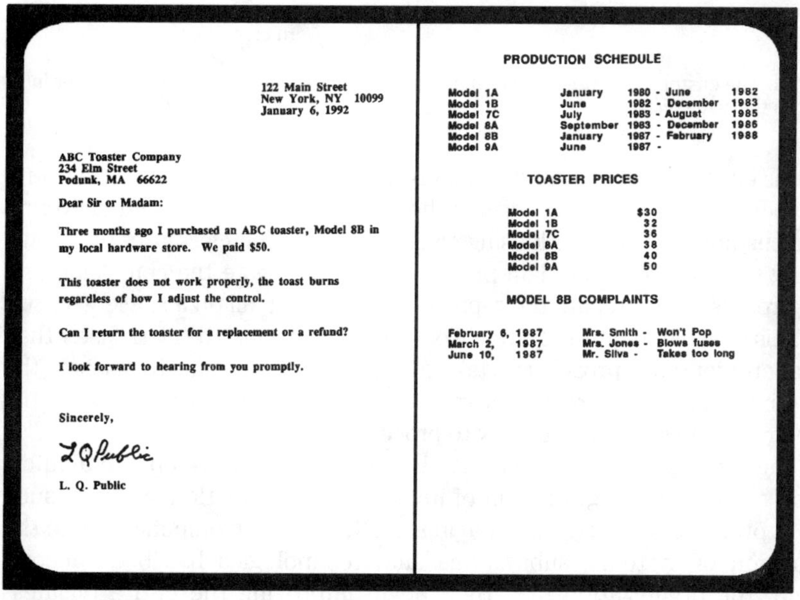

122 Main Street
New York, NY 10099
January 6, 1992

ABC Toaster Company
234 Elm Street
Podunk, MA 66622

Dear Sir or Madam:

Three months ago I purchased an ABC toaster, Model 8B in my local hardware store. We paid $50.

This toaster does not work properly, the toast burns regardless of how I adjust the control.

Can I return the toaster for a replacement or a refund?

I look forward to hearing from you promptly.

Sincerely,

L. Q. Public

PRODUCTION SCHEDULE

Model 1A	January 1980 - June	1982
Model 1B	June 1982 - December	1983
Model 7C	July 1983 - August	1985
Model 8A	September 1983 - December	1986
Model 8B	January 1987 - February	1988
Model 9A	June 1987 -	

TOASTER PRICES

Model 1A	$30
Model 1B	32
Model 7C	36
Model 8A	38
Model 8B	40
Model 9A	50

MODEL 8B COMPLAINTS

February 6, 1987	Mrs. Smith -	Won't Pop
March 2, 1987	Mrs. Jones -	Blows fuses
June 10, 1987	Mr. Silva -	Takes too long

Figure 7.2 Images of the customer's letter and related data from the mainframe on the same display.

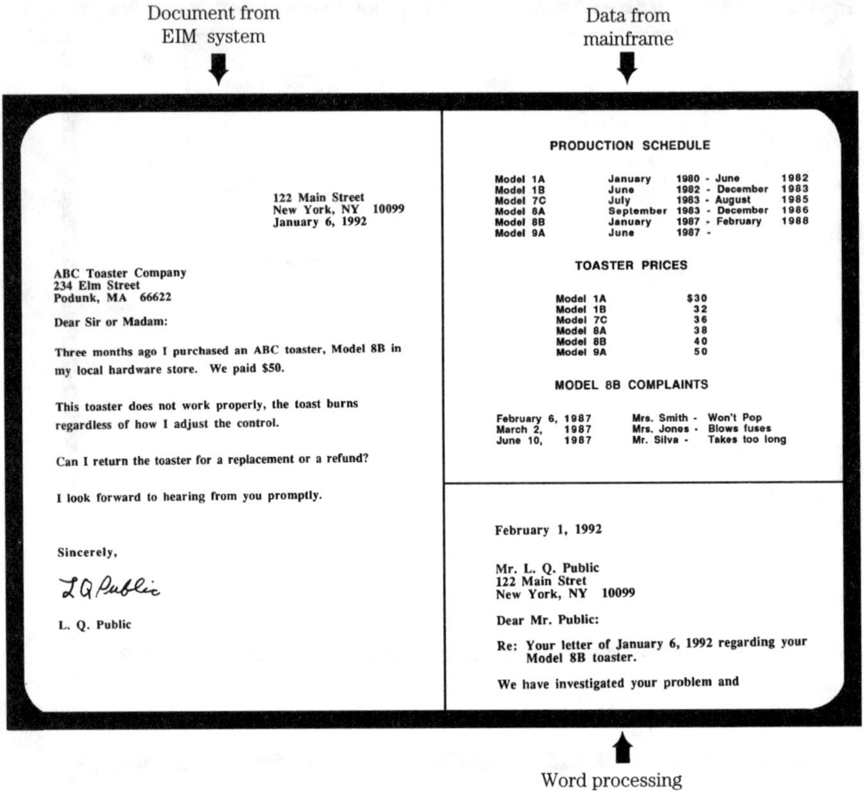

Figure 7.3 The customer's letter, data from the mainframe, and a word processing letter in reply, all on the same display.

In this integrated electronic information system, electronic imaging converts the paper letter to electronic form in the mailroom, and routes and delivers it to the customer service clerk using electronic mail. The clerk uses electronic imaging to view the letter, data processing to investigate the complaint, and word processing to prepare the reply to the customer, all on a single workstation. This type of integrated system would get the letter to the clerk faster than the paper-oriented approach, and take the clerk much less time to research the complaint and prepare a reply. The customer would get an earlier reply and the company should spend less money to process the reply.

A long-term goal of every organization should be to develop a total information system by integrating all of its separate information systems, such as data processing, electronic imaging, OCR, and electronic mail. Once the integration of systems, subsystems, and technologies has been accomplished, the organization may then begin improving the entire business process, most often called *workflow*.

Workflow

Most businesses today are going through what is being called *business process redesign (BPR)*, also called *business process reengineering*. BPR is a careful look at the way an entire organization does business, with the aim of substantially improving quality and productivity. Workflow techniques can play a major role here.

Workflow refers to the processing of documents throughout an organization. An integrated system automatically routes the flow of documents, in electronic form, from workstation to workstation throughout an organization. In this type of system, work usually gets processed faster on a LAN, where everyone can share documents and files. Several people can work with the same document or file at the same time.

In workflow electronic imaging systems, documents and files are not simply stored and retrieved, but are used to conduct the transactions of business. In a traditional paper system, a document passes from person to person, often with long delays between reviews. If the document is related to closing a sale or servicing a customer, the delay can be very costly. Suppose you substitute an electronic image for the paper document, scanning it when it arrives or when and where it originates, and forwarding it electronically to all sites simultaneously and instantly (see Figure 7.4). The sale is closed much sooner, the customer is satisfied immediately, and the organization might gain a valuable competitive edge.

It is important to distinguish between fully integrated, workflow imaging systems and strictly archival imaging systems, sometimes called "electronic file cabinets." An archival system uses imaging only for storing and retrieving documents. It replaces or supplements paper and microfilm records, providing more rapid access, better use of electronic technology, and more efficient storage. While beneficial, archival imaging is only the beginning of what imaging can do. It is best viewed as a starting point for the full use of electronic imaging technology.

Workflow imaging systems use software that creates an electronic "highway," speeding document images from workstation to workstation and priority to priority, automatically. They eliminate lost files, misfiles, and "out-of-files," increasing efficiency and productivity. Workflow is software-driven. Workflow software can automate information routing, task processing, and work-in-progress reporting. A simple routing system moves documents along a predefined set of processing steps. *Rules-based routing* allows normal and exceptional processing flows to move along a predefined path. *Conditional routing systems*, on the other hand, pass along normal information but will not process exceptional information. Task processing can vary from the very simple to the very complex. An insurance claims workflow program, for example, can cut checks, write cover letters, and record transactions in a database. Work-in-progress reporting tools can

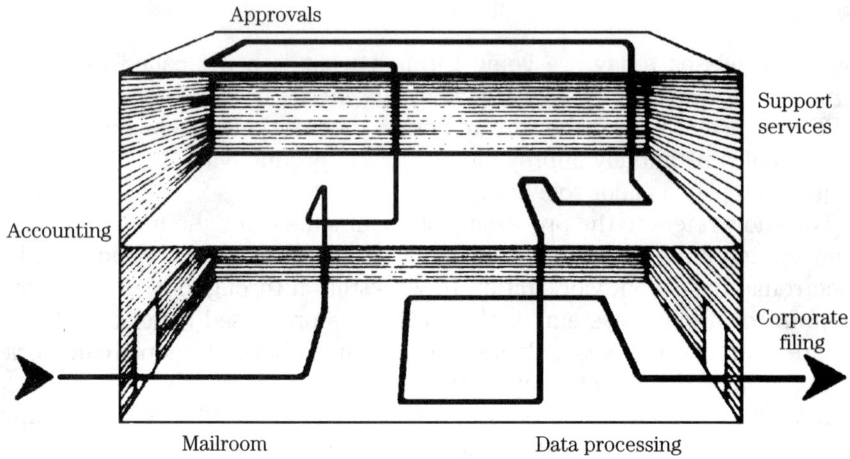

Figure 7.4 A workflow diagram.

tell you the location of outstanding items, explain which processing steps have been completed, and report whether processing has been on schedule.

Besides electronically controlling document images, workflow software automates a number of other management tasks, such as making productivity measurements, generating reports, and adjusting employee workloads and schedules. Consider time-sensitive documents for example. Workflow scripts can be designed so that each document has a specific timeframe for being processed at each workstation. If the document doesn't progress past a particular workstation within its assigned time, it's automatically routed to a supervisor for attention. This helps eliminate bottlenecks. It can also alert the supervisor that a particular user needs help, perhaps because he or she is new or has too heavy a workload. Such information can be invaluable when creating workflow charts.

Or suppose an invoice cannot be processed until the product arrives. No problem. Workflow software, with its rendezvous feature, suspends the document until the system is informed that the product has arrived. Then the system software melds the document with the new information and automatically dispatches the now-completed file to the accounts payable workstation. The entire process is fast, efficient, and simple.

Workflow scripts

Scripting tools are an essential part of workflow management. Workflow software allows users to write scripts, or programs, that detail where each document will go in the organization. Workflow scripts chart and control all documents that enter an organization. For each category of document, scripts can specify on which workstations a document must appear, and what other images should appear on the screen with it. Document processing is telescoped into a shorter period of time, so work gets done faster and more efficiently. Scripting tools eliminate "information float," that common condition in which vital documents are buried in in-baskets while people wait for them.

Workflow software's easy-to-understand menus and other configuration tools help users prioritize and schedule documents to be processed, so that the system automatically collects and distributes documents to electronic in-boxes in the proper order. As each person finishes processing a document, it is automatically sent to the next workstation, then the next. When processing is completed, there is no filing to be done. That was accomplished when the document was first scanned and indexed. All documents subsequently generated by workflow software—such as additions and annotations—are automatically filed in the appropriate file. In some cases, workflow configuration tools actually produce an on-screen flowchart. Some typical workflow products for various computer platforms are listed in Table 7.1.

One of the reasons workflow software is easy to understand is the use of electronic forms generators, which design forms for information capture. These generated forms contain intelligence fields that automatically prompt for processing actions and use scroll bars, radio buttons, and pull down menus to make processing even simpler. The same form, of course, can be used over and over again.

TABLE 7.1 Typical Workflow Software Products.

Product	Vendor	Platform
Action Workflow Management System	Action Technologies	Windows
EPIC/Workflow	Computer Technologies	Windows, OS/2, Presentation Manager
Process It	NCR	DOS, Windows, OS/2, Mac, UNIX
WorkFlo Business System	FileNet	UNIX, Windows

Workflow provides customized flexibility. Interfacing tools such as *APIs* (application programming interfaces) allow workflow processing programs to automatically open applications packages, such as spreadsheets and word processors, at appropriate points in the work process. No single program can fit every business need and environment, so workflow scripts can be created to match needs and operating procedures—for work files, menued procedures, image processing, mainframe interface, word processing, routing, and much more. Because workflow scripts can be quickly changed to reflect new ideas, procedures and needs, businesses are assured of continuing efficiency and productivity.

Additional benefits

Workflow software accelerates and simplifies all image processing. This helps new employees become productive earlier, and experienced employees become more efficient. Productivity improvements of 50% or greater are not uncommon.

Management benefits, too. By defining exact processing steps that are possible and required, management gains increased control over procedures. A methodology is created that employees can easily follow. As a result, quality improves and consistency increases.

In addition, workflow systems provide many more benefits:

- They automatically prompt the operator so procedures are not overlooked, and bring together required information in windows on the screen to speed processing.
- Multiwindow processing allows operators to suspend operations and handle interruptions.
- Duplicate keystrokes are eliminated by exchanging information between a mainframe and an integrated electronic imaging system.
- Management can check operator efficiency.

Standardization

Standardization is the process of formulating and applying guidelines for an orderly and efficient use of materials, components, and methods for the benefit and with the cooperation of all concerned. The result of a particular standardization effort, approved by a recognized authority such as ANSI (American National Standards Institute), ISO (International Standardization Organization), or CCITT (Consultative Committee for International Telephone and Telegraph) as described in chapter 8, might take the following form:

- A document containing a set of conditions to be fulfilled covering engineering and technical limitations and applications for items, materials, processes and methods
- A fundamental unit or physical constant, for example; ampere, meter, liter

The electronic imaging field needs guidelines and standards that will facilitate the following:

- The exchange of images between different systems
- The ability to buy components from a variety of vendors that work together as a total system
- The freedom of buyers not to be locked into a single source, forced to buy their next electronic imaging system from the previous vendor
- Competitive bidding for contracts to supply system upgrades and additional components
- Understanding between vendor and buyer

Types of standards

In the electronic imaging field, standards fall into four groups:

- Computer operating systems
- Communications
- Applications
- Testing and measuring

Not surprisingly, standards are a hotly debated issue in the electronic imaging business. Some observers, for example, contend that the lack of standards for formatting optical disks has slowed down the general acceptance of EIM. Others focus on the ongoing debate over standards for compressing and decompressing electronic images. Some critics even contend that the whole standardization issue is a red herring because some degree of standardization already exists (albeit within a single vendor's product line).

The fact remains, however, that no vendor today is selling an imaging system that is completely open. Customers, however, are pushing vendors in that direction. The trend seems to be toward creating industry standards that will ensure compatibility between hardware and software. This will make the job of integrating systems much easier.

The Association for Information and Image Management (AIIM) recently announced the creation of a Standards Strategic Plan. The plan grew out of a task force chaired by Michael Nier, standards development director of Eastman Kodak Company, and including eleven other members. This task force has created an extensive and detailed program for AIIM, focusing on achieving the following five goals over the next three to five years:

1. Establish leadership as a national standards developer for the application of document imaging technologies.

2. Be the generator of document imaging standards for the international community.
3. Be a major information resource concerning document imaging standards and related activities.
4. Continue to develop and accelerate the development of consensus-based standards.
5. Enhance AIIM's capability to manage, monitor, and direct the activities of its standard programs.

For more details on this committee's recommendations, see the sidebar on pages 111 through 115.

TWAIN and CALS

The joint development of *TWAIN*, a cross-applications platform interface program, meanwhile, could be a major step towards standardization. Essentially, TWAIN offers a common language that allows end-users to add images to a variety of documents without having to quit an application. TWAIN was developed by an ad hoc group of five industry leaders: Aldus, Caere, Eastman Kodak, Hewlett-Packard, and Logitech, with input from other imaging companies. The key element of TWAIN is the Source Manager (SM), whose primary role is to manage the connection between an applications and an image source.

Will TWAIN catch on? Most analysts agree that higher-end vendors will probably not be early TWAIN adapters. However, many analysts are hopeful that this well thought-out, practical solution will be adapted by vendors seeking to liberate applications and users from traditional device dependencies. Many people in the computer industry want to support TWAIN (which means buying TWAIN products) as a precedent for promoting continued cooperation between hardware and software vendors.

CALS (Computer-Aided Acquisition and Logistics Support) is another important facet of the move toward imaging standards. CALS, which began as a study by the Department of Defense in 1984, targets paper reduction. In 1990, the DOD stated that all its contractors needed to deliver documentation electronically following CALS guidelines. The second phase of CALS has increased the demand for optical storage media.

More about relevant standards and technical reports can be found in appendix C. Also, check with the AIIM technical department for any standards issued subsequent to the publishing of this book.

AIIM Standards

Goal 1: Establish leadership...

1. Establish recognition of AIIM as the preeminent standards-developing body for the application of document-imaging technologies. Document imaging technologies include the capture, storage, retrieval, integration, management, display, and distribution of document images. Forge alliances with public and private organizations that develop and use document-imaging information management systems, services, and technologies.
   - Educate those organizations that manufacture, use, install, sell, or otherwise deal with document-imaging technology about the importance and benefits of participating in the development and implementation of document imaging standards.
   - Promote outreach activities to increase the participation by members of selected vertical markets and consortia in the development and implementation of document imaging standards.
2. Develop and implement a mechanism to monitor and react to the impact of emerging technologies, including hardware, software, telecommunications, networks, storage media, and so forth, on the AIIM document-imaging standards program.
3. Monitor the adaptation of existing technologies in new applications such as the application of film, electronics, and high-density broadcast media for mass storage.

Goal 2: Be the generator of document-imaging standards

1. Achieve international recognition and adoption of ANSI/AIIM standards.
   - Introduce all ANSI/AIIM standards projects into the appropriate international program.
   - Raise AIIM member awareness of international standards issues and activities.
   - Use internationally accepted units of measurement in standards.
2. Adopt international standards as ANSI/AIIM standards.
   - Harmonize the content of appropriate international standards that should be introduced and adopted.
   - Continue to harmonize national and international standards adoption processes.

3. Define the benefits of standards to the international community, including
   - Compatibility of goods and services
   - Production efficiency
   - Improved product quality and reliability
   - Competitive prices
   - Steady product supply
4. Continue to evaluate the role of AIIM in certification and testing programs within the document imaging industry.
   - Monitor conformance and certification issues and activities.
   - Provide factual information regarding testing facilities and standards.

Goal 3: Be a major information resource

1. Promote the use of relevant standards within public and private information technology organizations, e.g., continue to develop and expand the imagery database.
2. Communicate with new national and international communities of users who affect or are affected by information technology standards.
3. Assure the inclusion of references to appropriate standards in *Inform* magazine and other AIIM publications by working with the authors or AIIM editors.
   - Submit articles to other information and image management publications.
   - Republish appropriate articles from other industry/vertical market publications.
4. Encourage the use of standards through an aggressive market strategy to include
   - Offering educational seminars
   - Communicating the benefits of standards using brochures, instructional videotapes, slide shows, and scripts for distribution, describing industry success stories
   - Describing potential cost savings
5. Devise communications tools to be used by standards developers in industry to reach their corporate management.
6. Develop easily accessible and highly publicized educational materials and programs to explain standards and encourage their adoption.
   - Continue with standards implementors' workshop for users, integrators, value-added resellers, consultants, middle management, etc., to promote AIIM standards and the development process.
   - Continue to provide an index to document imaging standards available from other standards developing organizations.

Goal 4: Continue to develop

1. Encourage greater participation in standards work.
   - Publicize organizations that use or promote standards.
   - Enhance the recognition program for volunteer participants.
   - Clarify the benefits to be gained from active involvement.
2. Establish, reconfirm, or restore liaisons in order to respond to the various needs of organizations and groups such as federal agencies, standards developers, certifiers, trade groups, and related industry users.
3. Expedite the standards process.
   - Review AIIM standards program policies and procedures to see how the process can be accelerated, for example:
     - Expedite the resolution of negative votes.
     - Increase the frequency and extend the duration of standards committee meetings.
     - Assign paid technical writers to committees to help draft relevant documents.
     - Allocate resources to support the fast-track process in accordance with recommendations from the Standards Board.
     - Assign the primary responsibility for each standards project to a committee member.
     - Review procedures used by other standards-developing organizations for any helpful or appropriate suggestions.
     - Gain approval from ANSI for any suggested process changes.
     - Increase support, visibility, and dedication of resources to the standards process.
     - Increase committee productivity by providing a typist for draft documents, meeting rosters, mailings, and document registers from the AIIM secretariat.
     - Increase AIIM staff productivity by providing electronic tools such as laptop computers and online mail services to facilitate the distribution and exchange of draft documents, meeting announcements, etc.

Goal 5: Enhance AIIM's capability

1. Coordinate AIIM standards programs with other standards organizations.
   - Minimize the duplication of effort on the part of standards-developing committees or organizations.
   - Generate and maintain a presence in related information technology communities for standards development and promotion.
   - Work with the appropriate ANSI boards, committees, and councils to coordinate work being performed by relevant standards developers.

2. Establish executive-level communications among standards groups, both among related committees within the AIIM standards program and among standards groups who affect or are affected by the document-imaging industry.
3. Establish a training and education program for AIIM standards committee participants, including the AIIM staff, committee chairs, officers, and technical editors. The subject matter should include the following:
   - Value of standards
   - Legal issues
   - ANSI/AIIM policies and procedures
   - Effective meeting leadership
   - Preparation of minutes and reports
4. Establish, reinstate, or strengthen active liaisons with appropriate vertical markets as identified and prioritized in the AIIM industry study:
   - Banking and finance
   - Healthcare
   - Insurance
   - Government agencies
   - Manufacturing
5. Establish, reinstate, or strengthen active liaisons with other appropriate industry organizations and user groups:
   - Agency Company Organization Research Development
   - American Association of Motor Vehicle Administrations
   - Association of Information Managers
   - Automated Records Managers Association
   - Commission on Preservation and Access
   - Data Processing Management Association
   - Digital Image Applications Group
   - Equal Partner Implementation Committee
   - National Association of Government Archivists and Records Administrators
   - Research Libraries Group
   - Society of American Archivists
   - Society for Imaging Science and Technology
   - Special Interest Group for CD-ROM Applications and Technology
   - Tape to Head Interface Committee
6. Establish, reinstate, or strengthen liaisons with other standards developers, as follows:
   - American Library Association
   - American Society for Information Sciences
   - International Business Forms Institute

- National Association of Photographic Manufacturers
- National Information Standards Organization
- National Printing Equipment and Supply
- Records Managers Association
- Nuclear Information
- Special Libraries Association
- Telecommunications Industry Association
- X3 Information Processing Systems (e.g., X3BI1, X3H3, X3L3, X3VI, and X3WI)

7. Establish a strategic plan to encompass the current and future directions of the AIIM standards program.
   - Established goals should remain stable, but objectives should change as time and events and alternative strategies require.
   - The document will be reviewed annually by the AIIM Standards Board.

Legality

One of the most pressing legal issues facing the imaging field is, "Is information stored on optical disks admissible in a court of law?" The answer is "Yes, but . . ."

In the United States there are seven different types of evidentiary laws, more than 50 different judicial jurisdictions, and more than 100 regulatory agencies. Thus, an answer to this question is not simple.

All 50 states have laws that provide a solid legal foundation for admitting records stored on optical disks in court. Although the specific legal basis might differ from one state to another, in most states it is found in the Rules of Evidence or the statutory laws. The key element of the Rules of Evidence provides for the admissibility of data compilations. This concept is included in the Uniform Rules of Evidence and Federal Rules of Evidence. The concept of a data compilation has been used to admit magnetically stored information, and therefore it should be applicable to records stored on optical disks.

A *data compilation* is information in machine-readable coded form. Computers typically use ASCII code, which is a pattern of eight zeros and ones for each letter of the alphabet and number. Electronic imaging systems also use zeros and ones to represent information, but they represent light and dark points of the image as described in chapter 2. The Federal Rules of Evidence and Uniform Rules of Evidence specifically authorizes the admissibility of a "data compilation, in any form" that has been kept in the regular course of business.

There are two other important factors regarding the admittance of information stored on optical disks: the Best Evidence Rule and the admissibility of hearsay. The hearsay issue is raised when it is not possible to cross-examine evidence. The issue in submitting a document is how do you cross-examine a document? This situation can be overcome by the Business Records Exception to the Hearsay Rule, which recognizes that certain records created and relied on in the regular course of business have a certain circumstantial probability of trustworthiness, and therefore ought to be received in evidence. The Business Records Exception to the Hearsay Rule essentially says, "If the record is good enough for a business to rely on in the normal course of its activities, it is reasonable to expect that it is good enough for the courts to rely on in their pursuit of the truth and, in turn, use to decide about a matter pertaining to the business."

The Best Evidence Rule states that the original should be received as evidence unless a satisfactory reason is presented for unavailability. The option for providing copies of records is addressed in the Federal Business Records Act and its state counterpart, the Uniform Photographic Copies of Business and Public Records as Evidence Act (UPA). These laws permit the admissibility of any record that has been "kept in the regular course of business and copied or reproduced by . . . any photographic, photostatic, microfilm, microcard, miniature photographic, or other process which accurately reproduces or forms a durable medium for reproducing the original." The process of recording information using binary optical code clearly falls within the law's language of "other process which accurately reproduces or forms a durable medium for reproducing the original."

Paper records may therefore be destroyed after the information has been recorded on optical disks. The Uniform Photographic Copies of Business and Public Records as Evidence Act states that copies created in the regular course of business are "as admissible in evidence as the original itself in any judicial or administrative proceeding, regardless of whether the original is in existence or not." Further, it states that "the original may be destroyed in the regular course of business unless held in a custodial or fiduciary capacity or unless its preservation is required by law."

Optical disks do not have to be archival or have any minimum quality level, such as a specific number of pixels, to be acceptable as evidence. Therefore, legality, archivability, and quality requirements are separate issues.

It is very important that optical disk users differentiate between the legality of the media and the legality of their system. Just because there is a solid legal foundation for admitting optical disks, does not guarantee that your electronic imaging system will be acceptable. In legal proceedings, it is usually necessary to show that the specific system from which the documentary evidence has been produced has the capability of producing accurate, reliable, and trustworthy documents.

Copyright is also an important issue. No one denies that the potential for abuse exists. Drawing and redrawing the line between fair use of copyrighted material and infringement could require large-scale legal attention.

Of course, the basic information on legality presented here is just guidance, not for legal advice. For that, you should consult your own legal counsel. (The definitive reference on the subject of EIM legal issues is *Legality of Optical Storage* by Robert F. Williams.)

Chapter 8

International Issues

Electronic imaging is truly an international business. Many of the companies leading the imaging field are international in scope, including its vendors, consultants, and implementation pioneers.

Conglomerates

Many of the largest companies involved in the electronic imaging and printing industry are international conglomerates. For example, the Wace Group is the largest color prepress group in the world. It has more than 100 operations in the U.K., U.S., and continental Europe. Ian Laurence, Wace's marketing director, has pointed out that cost-cutting measures by clients mean that trade shops have to get bigger to survive, to be able to handle work on a global basis.

In many cases, success stories in one country end up being duplicated in other countries. For example, the art director at USI, Inc. uses a prepress electronic imaging system that can cut in half both the time and the money needed to produce a color catalog. CyberChrome, Inc. is the color prepress production value-added reseller (VAR) providing the system for the Connecticut firm. (Prepress technology advancements give users the capability to transmit a publication to a printing house.) Two other VARs in the U.S., Lewis Printing Company and Wintech, Inc., and two in Europe, Tint Publishing AB and BlackBit DTP Studio, soon signed agreements to carry the CyberChrome products.

In general, however, American companies have a history that ignored internationalism. Business professors point out that for many years American companies avoided aggressive international marketing. These experts explain that American companies preferred simpler, safer domestic markets to foreign companies. Foreign trade, after all, involves new languages, different currencies, legal and political uncertainties, and adapting products to different customer needs and expectations.

However, within the last decade, American companies find themselves increasingly drawn into the arena of international marketing. Slower growth in the domestic GNP have forced companies to search for new opportunities. Higher manufacturing and research and development costs, especially with high-tech companies, force many to "go global" to recover costs.

Many American companies have discovered the fact that international markets offer a variety of opportunities, and that they don't have to abandon domestic markets to take advantage of these new opportunities. Indeed, they are finding other markets attractive even after the extra costs and encumbrances of operating abroad are tallied up.

Electronic imaging becomes an important part of the international growth for many companies. For example, in 15 years, Alexander and Alexander had grown from a U.S.-only insurance brokerage to a worldwide operation spanning 70 countries, with 16,000 employees. Revenues have increased from about $40 million in the middle 1970s to $1.2 billion in the 1980s.

The insurance business, of course, is very paper-intensive. Alexander and Alexander turned to an electronic imaging system with optical disks. Each disk could store 14,400 highly complex documents. Eventually, the system expanded to where a jukebox made sense.

Meanwhile, international communications technologies have forced cooperation between international trading partners. By the middle of the 1980s, pressure from European trading partners eager to eliminate paper transactions forced some American companies to set up electronic data interchange (EDI) networks linking Europe and the United States. Despite the difficulties of building expensive and complex networks complicated by regulations and fee structures that vary from country to country, MIS executives have insisted that the investment more than paid for itself in increased efficiency.

Chrysler Corporation set up an EDI system linking European operations as early as 1988. Rockwell International, however, was one of the first companies to link the United States and Europe. A pioneering project in 1989 began by creating the first leg of a European EDI network in France. Rockwell's $2.15 billion automotive division, based in Troy, Michigan, supplied parts to European manufacturers, including the Renault Group and Peugeot SA. EDI was introduced to play an important role in establishing just-in-time manufacturing operations. Plans call for immediate expansion of the network into Germany and other countries.

The United Kingdom is serving as the central hub for the network. A Rockwell spokesperson reports that it was 25 percent cheaper to route overseas EDI communication through London than to run a direct line between France and the United States. Dedicated lines now link London headquarters to a front-end processor in Paris, while public lines link customers, suppliers, and manufacturing sites.

Doing Business in Europe

England seems to be an especially fertile market for electronic imaging. BIS Strategic Decisions has estimated that the United Kingdom's electronic imaging market will grow by 60 percent per year between 1991 and 1995. The prediction is that the current £24 million market will reach more than £170 million by 1994.

Filenet, for example, set up a U.K. subsidiary in 1991. Prior to that, Filenet marketed in England through a partnership with Olivetti. Their marketing plans have called for expansion into areas such as government, utilities, pharmaceutical, publishing, and manufacturing as well as traditional markets such as finance and insurance. One of Filenet's largest customer is National Westminister Bank, which uses more than 100 workstations in its credit card customer department.

Britain has also championed electronic imaging in its international operations. British Petroleum (BP) in Norway, for example, has installed an advanced drawing system using optical disks. The system transmits drawings from plants in Norway and Great Britain. Finished drawings are then beamed via satellite to an oil rig in the North Sea.

BP estimates that the project will generate some 35,000 engineering drawings, entered into the system in conventional form. BP then converts them to a common electronic format, permitting transmission from the main site at Tanager in Norway to fabrication sites in Norway and England before heading to the new oil rig. BP also incorporated an unusual requirement into the system design: All documents must be destroyed after they were scanned. Planners looked for a format that would accomplish the following:

- Speed up the supplying of drawings
- Eliminate mailing costs
- Reduce space requirements

Spectrum Datagraphics Systems International, a British firm with customers in France, Norway, Holland, Italy, and the Middle East, created the system that would bitmap all documents and store them on WORM optical disks. The core of the Tanager system is a Sun 3/280 computer acting as the file server, fitted with three magnetic disks and two 12-inch WORM disks. Other equipment includes the following:

- A 200 dpi scanner workstation
- A Versatec 7236 AO electrostatic plotter
- Six viewing/editing workstations

The total value of the Spectrum Datagraphics contract was £750,000. On the other hand, the total budget for the oil project was £750 million. BP, Statoil, and Conocco of Norway, are targeting as much as 60,000 barrels a day of oil and gas liquids.

As large as the European market is, doing business there, especially in the electronic imaging arena, has its quirks. Alan Ryan, writing in *ComputerWorld* (December 24/January 1, 1991), saw the following business functions being of primary concern to MIS managers expanding in Europe:

- Distribution and inventory control
- Marketing and sales
- Manufacturing
- Data protection
- Human relations

Despite the spirit of cooperation of the European Economic Community (EEC) and the agreements which went into effect in December 1992, serious cultural and political differences exist between European countries. A significant challenge for MIS departments, for example, is to convert sales commissions into different currencies. Because currency exchange rates change daily, this process can prove extremely difficult. Countries in Europe also maintain different measurement systems. The United States and Great Britain use the British system of measurement (pounds and inches), while most EEC countries prefer metric measurements. Privacy issues are also sure to arise among countries interested in protecting their own citizens from scrutiny by other powers.

Human resource issues are proving especially thorny. Europeans, many cultural experts observe, are adverse to moving to a new country. Centralization, however, means that either old workers must move or new ones be hired in their place. Employees crossing international boundaries also have to contend with differences in retirement plans, medical benefits, and other compensation. Companies might end up having to create new departments and job classifications to cope with transborder transactions.

Security is a major issue underlying all internal computer communications. The harsh reality is that with the great proliferation of data communications both in the United States and abroad, the potential for invasion of privacy, security leaks, and other breeches of confidentiality is great. More than 20 countries have passed laws to deal with problems when information flows across national boundaries. Transborder information flow, then, is governed by legislation at the national level. Spain, for example, requires that money be deposited in escrow before files can be transmitted electronically to another country. Sweden has created a government agency to approve any transmission of personal data outside the country.

The Middle East

Electronic imaging sales have expanded rapidly in recent years for the Mideast. Much of this expansion can be attributed to joint ventures in the private sector. The Network Connection, of Georgia, has signed contracts with Al-Othman and Al-Saraff Computer Services, Kuwait City resellers, for distribution of the Triumph image processing system (TIPS) throughout the Gulf States. Meanwhile, other Americans are working with governments in the Mideast.

For example, Kofax Image Products, in California, has signed a contract with the Saudi Arabian Monetary Agency to supply their imaging software and hardware for a network that handles all of that country's fiscal transactions at Ridyah and ten branch offices.

The Far East

Laserdata, Inc., one of the leading U.S. manufacturers of PC/LAN-based image processing systems, signed a $4.5 million joint venture agreement with Kyocera, a $3 billion Japanese manufacturer of electronic components and computer equipment. Kyocera received technological expertise and equity in Laserdata. Laserdata gained additional clout. The 10-year agreement states that Kyocera will sell Laserdata hardware and software in Japan and other international markets. The agreement also gives Kyocera the right to incorporate Laserdata imaging technology into future Kyocera products.

Latin America

Many economists are looking at South and Central America as one of the growth markets for the 1990s. Of all the Latin American countries, however, Brazil seems to be leading the way in the development of the imaging industry. Canon, Eastman Kodak, IBM, Wang, Hewlett-Packard, Laserdata, 3M Unisys, and several smaller companies are all represented in Brazil. In addition, Sao Paulo hosted Brazil's first imaging conference in 1992. The show is the second largest imaging showcase in the world and offers more than 60 sessions in both English and Portuguese.

Canada

Canada is also using electronic imaging as a tool for its growing transnationalism. For example, attorneys in Canada use electronic imaging to keep up with the latest decisions. Today, many Canadian lawyers turn to QL Systems Ltd., of Ontario, for the latest, most accurate information available online 24 hours a day. As quickly as courts make a decision, they let QL Systems know. The information is available online to customers within the hour; if a court decides today on some legal issue in Toronto, the information will be available to a lawyer in Vancouver within hours.

To handle the heavy workflow of paper generated every day, the company uses four highly advanced document input systems by Xerox Imaging Systems (XIS). The document input systems consist of XIS-proprietary intelligent character recognition software integrated with heavy-duty, high-speed XIS K5200 scanners equipped with automatic document feeders. The K5200 scanner can scan 3,000 characters in 10 to 17 seconds including processing time. The recognition software automatically learns new typefaces. There is also a text-verification feature to allow the user to optionally interact with the system to provide the highest degree of accuracy. The K5200 also recognizes characters from several foreign languages, including

- Danish
- Dutch
- English
- Finnish
- French
- German
- Italian
- Norwegian
- Spanish
- Swedish

A staff of 85 people operate the four K5200 systems using two shifts, totalling 16 hours a day. As soon as they receive a document, it is run through a K5200, which reads the data and converts it into a WordPerfect file. The file is then placed into one of their databases.

Supporters of the system point out that the system saves both time and money. It also has a democratizing effect. With the help of the K5200s, QL Systems gives a one-person law firm the same access to timely information as a large firm at a cost of less than five dollars for the retrieval of several cases.

Eastern Europe

Eastern Europe is also a promising market. The broad changes of the past few years have altered the political realities of this region, leaving many businesspeople wondering how they can take advantage of the massive new market. As early as January 1991, the *Seybold Report on Publishing Systems* observed:

> "Crumbling Barriers" could have been the theme of the 1990 Seybold European Publishing Conference, held December 3-5, 1990, at the Netherlands Congress

> Center in the Hague, Holland. That message seemed to appear in nearly every presentation, every session. People everywhere talked about the impending demise of barriers to trade, the crumbling barriers between users and vendors, and the reduction in barriers between people in general. The speakers who didn't address these issues directly at least mentioned them in some fashion.

Hungary and Czechoslovakia joined the conference for the first time, opening up participation to Eastern bloc countries. Hungary was represented by Computer Media, a system vendor and integrator established in February 1989. It started as the exclusive local distributor of Ventura Publisher, which it customized (including the Gem environment) for the local market. In addition, Computer Media has assembled a few publishing system modules using the exclusive Hungarian distributor of XyWrite and a VAR for Linotype and Dell Computer. It also is the exclusive distributor for Etap monitors. It worked on the first CD-ROM application in Hungary, a record of the Hungarian Rules of Law in force.

Analysts point out that the Eastern European market has missed generations of development and is suffering from a serious lack of equipment, but that it has an excellent level of user awareness. Statistics show that the number of publications in this market are at least 210 newspapers and more than 5,000 magazines. The desktop publishing market, meanwhile, has been compared to the elusive Yeti (the mythological Himalayan "snow monster"): no one has seen it yet, but it is bound to be big.

As promising as doing business with the Eastern bloc is, the prospect is not without risks and drawbacks. The problems of doing business with the Eastern bloc include the following:

- Frequent currency fluctuations causing payment problems
- Appalling telecommunications
- A lack of mainframes to back up desktop computers
- A huge licensing and copyright problem
- No perceived need to buy multiple copies of software, since indiscriminate copying is the norm
- Low productivity
- Weak marketing research available

Russia and other former Soviet republics might also form a promising market. Gone are the days when those republics did little to recognize the development of intellectual property such as software and integrations system. In the old days, industry analysts observed, people had little incentive to create software or systems. People would create code, but their products would languish because no one advertised it, improved on it, resold it, supported it, or wrote manuals.

International Not-for-Profits

Profit-making ventures are not the only organizations taking advantage of electronic imaging on an international basis. For example, the Library of International Relations (LIR), in Chicago, is using an imaging system not only to preserve valuable and rare documents, but also to make them more accessible worldwide. In 1991, LIR initiated a pilot project to scan and store electronically the images of documents that could be transmitted on request to libraries, businesses, and other institutions worldwide. The library has accumulated a 400,000-page database. Historians are especially interested in seeing the LIR's extensive library of League of Nations documents preserved this way.

International Standards

One important reflection of the move towards a global market is the emergence of international standards in the computer business. Standards are often set by an interlocking force of national and multinational organizations. National organizations, for example, include the American National Standards Institute (ANSI) and the Institute of Electrical and Electronic Engineers (IEEE). International organizations include the Consultative Committee for International Telephony and Telegraphy (CCITT) and the International Standardization Organization (ISO).

ANSI

The function of the American National Standards Institute is to oversee voluntary standards work in the United States. The organization works with both the private and public sectors. ANSI does not develop standards of its own; instead, it

- Identifies areas in which standards are needed
- Ensures that the standards developers (such as the Association for Information and Image Management) and committees working under ANSI's auspices follow due process when developing and approving proposed standards as American National Standards
- Publishes a catalog of standards
- Publishes newsletters, reports, reviews of standards, development of policies and procedures, and the sale of national and international standards

The accredit committee, X12, is one of the ANSI committees most commonly associated with standards for electronic data interchange (EDI), a technology developed to facilitate the exchange of business transaction documents such as invoices and freight bills. In turn, the X12 committee also coordinates the work of the X3 committee, which is responsible for information technology standards. It also works with the TI committee, formed after the

1984 divestiture of AT&T to develop consensus communications standards that were formerly the responsibility of AT&T.

The IEEE

The IEEE professional society is a prominent example of an ANSI-accredited standards organization. Two IEEE groups focus on the development of standards for the communications industry and information processing:

- The Computer Society of the IEEE
- The IEEE Communications Society

The Computer Society of the IEEE began work in 1980. It produces most of the organization's information systems standards, focusing primarily on electronic devices. On the other hand, the IEEE Communications Society works in the area of communications-related standards. Both groups, in turn, coordinate their activities with the ANSI-accredited X3 and TI standards committees.

The CCITT

Headquartered in Geneva, Switzerland, the CCITT is an agency of the International Telecommunications Union (ITU). The CCITT has developed standards covering telephone transmissions and data transmission. Many of these standards are published as regulations so that they will be mandatory for manufacturers. In addition, the CCITT publishes recommendations that often end up as the basis for international agreements. Typical recommendations cover such areas as

- Technical issues
- Administrative issues
- Tariffs
- Terminology

The CCITT uses study groups to accomplish its goals. For example, Study Group VII covers data communications networks. Study Group XVII covers transmission over the telephone.

The ISO

The International Standardization Organization, established in 1947, is also headquartered in Geneva. Its mandate is broad. The organization seeks to develop consensus on international standards. Their overall objective is to improve international collaboration through communication in order to facilitate international trade growth, smoothly and fairly. Individual countries send their own representatives to the ISO. The U.S. representative to the ISO is ANSI.

One of the most important contributions by ISO to the new communications standards for electronic communication is the Open Systems Interconnection (OSI) model. The ISO created the OSI model, in part, to help put existing standards in perspective by defining their position within a comprehensive communications system.

The OSI model has seven layers. Each of these layers provides a segment of the overall communications function required for distributed applications. In turn, standards can be developed for each layer. The layers are as follows:

7 Application
6 Presentation
5 Session
4 Transport
3 Network
2 Data link
1 Physical

Changes made in one layer need not require changes to existing software in another layer. This makes it possible to adapt the model to emerging communications standards and continuous modifications caused by technological changes.

Individual governments end up promoting OSI standards. Government Open Systems Interconnect Profile (GOSIP) organizations throughout the world seek to influence these changes. The United Kingdom GOSIP committee, for example, has been a leader in this movement and often is held up as a model. The major outcome of GOSIP efforts in the U.K. has been the government OSI procurement standard. This standard requires all federal procurement of communications hardware and software to conform to OSI-based standards. Waiver can be granted only after the need for a waiver can be convincingly demonstrated.

The United States government, meanwhile, also supports the TCP/IP, or Transmission Control Protocol (TCP) and Internet Protocol (IP). These standards were developed by The Department of Defense in the 1970s as part of the research for the ARPAnet wide-area communications network. Although it was originally a noncommercial standard designed for use by computers communicating between universities and government laboratories, TCP/IP soon gained commercial acceptance because of the following factors:

- Businesses needed to show TCP/IP support in order to obtain government grants.
- Businesses needed to connect to government networks.
- Prior to the acceptance of OSI standards, the government required TCP/IP compatibility as a condition for government contracts.

Chapter

9

Evaluating Imaging Needs

Electronic imaging systems do not exist in a vacuum. Unless you are working for a VAR or computer company that specializes in electronic imaging, odds are that electronic imaging is a support function. As such, your electronic imaging system should be designed to offer solutions to organizational, technological, and project-level problems. Whether you are working within a small workgroup or implementing sweeping organizational changes on the departmental level, your electronic imaging project should set technological goals that fall in line with your organization's goals.

The most common question emerging in the imaging industry is no longer, "Why should a system be implemented?" but "When and how should it be implemented?" Despite occasional scattered resistance, the experience of many organizations over the past five or six years has reinforced the theory that imaging does offer competitive advantages.

When digital storage and communication of images first appeared on the scene, the vision of an all-pervasive, fully integrated system was considered the logical use for these technologies. The buzzword of the 1980s was "enterprise-wide." Forecasters led people to believe that an invisible hand would reach down from the Electronic Imaging Olympus and mold an office automation revolution of apocalyptic proportions.

These days, however, electronic imaging appears to be taking a new turn toward smaller units. While organizational battles are still waged in some boardrooms, the rise of "guerrilla wars" inside smaller work units is on the rise. In many organizations, the immediate needs of smaller work units are taking priority over those of the organization as a whole. Departmental-level solutions are becoming more common.

The challenges facing the would-be champion of electronic imaging have less to do with technological solutions than with economic and political ones. In economically unstable times, the failure to present clear-cut examples of where electronic imaging has triumphed and a preponderance of unrealistic expectations has placed imaging customers in a tricky position. Organizations are demanding more than a solution to the file cabinet problem; they've moved beyond the point where all they want is a "file and retrieve" system.

Ultimately, better communication between the buyer and the sellers is needed for electronic imaging to thrive. Vendors have the responsibility to prepare senior management to acquire the technology. Otherwise, the buys will not take place. On the other hand, purchasing organizations require a better understanding of their own needs so that they can become full partners in the electronic imaging process. They can start with goal setting.

Organizational Goals

The best way to look at any electronic imaging system is the same way you look at any other management tool—its job is to further the organization's goal. The first step in determining your organization's requirements for an imaging system, then, is to understand what, exactly, your organization is trying to do. In the jargon of business management, you are looking to establish *goal congruence*. The electronic imaging system then should serve to further your organizational goals.

A word of warning, however: The process of clarifying organizational goals is no simple task. Experts warn that it is likely to be time-consuming. If your organization does not have an up-to-date strategic organizational plan, strategic planning might be needed. The development of a strategic plan requires a thorough review and analysis of an organization's activities on all levels within that organization. This process can take several months, or even years, to complete.

No one would attempt to build a large office building without drawing blueprints first. Wouldn't it be terrible if the fifth floor of a skyscraper did not line up in some way with either the fourth floor or the sixth floor? Strategic planning is the blueprint-drawing process, so that all levels of an organization fit together. Strategic planning is made more difficult because you usually have to draw blueprints for an organizational "building" that has already been built.

Certainly, such a massive project is far beyond the scope of this book or your computer imaging efforts. (And you thought computer people could do everything!) However, as a rule-of-thumb, you need a well-prepared statement of organizational goals in order to determine the best way to support those goals, especially when you're dealing with "big picture" reorganizations that cut across departmental, and even company, lines.

The payoff of implementing or refining an electronic imaging system is not only reduced costs, although this is certainly an important benefit. Often, an organization's first imaging system serves as the testing ground for its ability to set and meet goals. In other words, electronic imaging has become a strategic planning tool, which is why senior management needs to become involved.

If your senior management doesn't want to "be in the loop," watch out! This could be a clear sign that there is danger ahead for your electronic imaging project. For electronic imaging to work effectively, senior management should be interested not only in ground-up planning but also in review. Failure to secure senior management support will dwarf the impact of an electronic imaging policy. If they simply treat electronic imaging like a wind-up toy—putting it in motion and then getting out of the way—others might soon lose interest in the process. Without full support from top management, electronic imaging champions are likely to come up against management unprepared (or unwilling) to make a decision.

To really work, the implementation of electronic imaging policy requires massive workforce support—unfortunately, office politics comes into play. Every individual within a company is a prospective friend or enemy of an electronic imaging project or policy. While that's true, of course, of anything that causes an organization to change, electronic imaging projects are particularly vulnerable. These high-profile changes run the risk of becoming political dynamite because so many people are involved over such a long period of time. And remember, everyone still has to do his or her "real job" during the changeover.

Your organization might need outside consultants to help with the strategic implementation of electronic imaging. Indeed, even if you don't need outside help to do anything else with electronic imaging, the strategic planning phase is often the time to use a consultant's expertise. If your goal is to find an impartial consultant for strategic matters (or any other reasons), however, your search might prove disappointing. Consultants tend to be tied to specific vendors, simply because the technology changes so quickly that certain vendors inevitably know more about certain technologies than anyone else. Consultants who pride themselves on integrity end up working more closely with vendors most capable of solving their hardware integration and software application problems. In all fairness, in many cases, managers just tell consultants, "Bring us something that works," or words to that effect. Consultants want to be able to deliver solutions with as few problems and as quickly as possible.

Problems, however, can arise when consultants forget that electronic imaging is just one technological tool used by an organization. The quick and easy electronic imaging solution might clash with an organization's needs for continuity. Fortunately, vendors have begun to understand that "continuity" is not just another business buzzword. Customers are really serious when they demand the ability to integrate electronic imaging systems seamlessly with other technologies. Sometimes, moreover, companies themselves end up changing to take better advantage of electronic imaging. Many, for example, end up reeingineering workflow patterns to take full advantage of electronic imaging technology.

A word of warning, however, from George Hall in the April 1992 issue of *Inform*: the strategic-advantage argument is often oversimplified. In order to solve that problem, he suggests customers consider a hierarchy of advantages in strategic goals:

- Cost reduction
- Upgrade leverage
- Entrepreneurial leverage

Technological Goals

If you're not a technologist, you might be thinking, "Why do I need to have technological goals?" The world of high technology changes very rapidly; you need to come to grips with these changes before your organization makes a significant change in its installations. Before setting goals, your organization needs to review the image industry's products, supporting technologies, and markets. This informed vision will help you develop a technological strategy to reach your organizational goals.

The next step, then, is to draw up a statement of technology goals. Such a statement need not be lengthy or impressive-looking. In setting your goals, remember that trade-offs are often necessary. For example, if resolution is your top technological priority, your goal might be to find an ultrahigh resolution scanner. With that goal in mind, you would comb the trade publications and the trade shows looking for the best resolution available. On the other hand, your organization might not need hardcopy output very often. In that case, you won't need to focus very much on printer questions. Your statement of technological goals should reflect these priorities.

Your statement should not be too detailed, however. The idea is to have a statement capable of guiding your longer-range decisions about systems functions. It must be based on a solid understanding of what imaging is and what it can do. Get your MIS staff involved in this educational process of understanding and accessing technology by asking for their input. Their findings—bibliographies, market studies, vendor surveys, and other valuable materials—will prove an invaluable resource. Also, this researching is a good way to empower these people, who must work most closely with the new system.

Of course, the electronic imaging environment is constantly changing. Your technological plans should reflect this reality. The good news, as your MIS staff should be the first to tell you, is that prices are dropping, while performance is on the upswing and should continue to improve as more beneficial applications and systems continue to appear. Outside consultants or MIS staffers can add expertise in guiding your organization along the evolutionary path. That means your electronic imaging program will take place in stages. Remember, as a child, when you were given clothing a little bit too large so that you'd "grow into it?" Creating a plan with stages in it gives your organization a chance to grow along with new technologies.

How much growth should you build into your imaging program? Conventional estimates for the implementation of an imaging system suggest the following time blocks:

- Six to nine months of initial education in technology
- Three to six months for the system's champion to surface and to figure out which senior partner will support that champion
- Three months to write the RFP
- Three months to get it through the acquisition process
- Three to six months of effort until the system is finally turned on

Project Goals

In setting project goals, all of the hard work you put into setting organizational and technological goals comes into practice. Project goals are more brief than either statements of organizational goals or technological goals. They can and should be standardized to include some or all of these three general categories:

- Operator goals focus on scanning and indexing documents and filing them as computer images.
- Delivery goals, such as design, testing, documentation, and functional performance, concentrate on getting the images to the desktop facilities where they can be put to use.
- Integration goals require the development of collaborative applications and the integrating of images into existing applications.

Because you will be setting project goals repeatedly, your organization should consider creating a standardized project goal form. This form will enable you to make fewer errors and to move documents through your electronic system quicker. Have these forms designed and tested before you start conversions.

Project goal forms should be on 8½×11-inch sheets of paper. Each type of form should be given an identifying code, preferably a numeric one. Put the form identifier at the top of the form because that's the best spot for the scanning process. Any ink on your blank form will take time to scan; therefore, to be as efficient as possible, consider the following tips:

- Use bar coding if possible.
- Don't label the blank slots on the form "Name," "Address," or anything else. OCR will take the time to read your labels as well as what's filled in the blanks.
- Make the form as white as possible.

Strategic Planning

Strategic planning is the systematic development of a plan to employ the economic forces of an organization for maximum support of adopted policies. MIS systems are an integral part of that planning. The modern concept of strategic planning for MIS is the process of systematically thinking through a business situation, anticipating the changes that data suggests, and developing a series of strategic options or alternative objectives to guide management.

Strategic planning is based on the resolution of management concerns or issues. Strategic analysis, the process of thinking through these issues, answers the question, "Are these objectives valid?" In other words, do they do what they are supposed to do? To answer this question, objectives or options are subjected to a series of validity tests. Once they clear that hurdle and validity has been established, the objectives may be supported with a plan known as an *implementation action program.*

Strategic planning should never be confused with long-range operational planning. Strategic planning has little, if any, relationship to time. The best way to think of strategic is as planning in response to corporate business initiative. In other words, this is the "big picture" that college professors like to discuss. The *strategy* of *strategic planning* refers to the strategy of the organization as a whole.

Strategic planning focuses solely on solving critical problems. It might involve forecasting over a long period, a medium period, or a very short period. The driving force, however, is always the need to solve a crucial problem as quickly as possible. The purpose of incorporating MIS into strategic planning is to link business and computer strategies. Organizations do this because they identify areas of change and achieve this change through the use of informational technology.

Operational Planning

Operational planning is the development of a summary of MIS services over a particular time period or for a particular project. This summary identifies opportunity areas that have been analyzed by systems personnel and lists both potential projects and projects under development. In most organizations, many more projects are considered and analyzed in the operational planning process than will be finally approved for implementation. Operational planning also evaluates and ranks projects and services by priority and translates the results into anticipated schedules, resource requirements, and plans or action.

Projects considered in operational planning arise from various sources, including the following:

- Routine requests from users
- Technologically desirable changes within MIS
- Routine maintenance of existing systems

Some operational planning is long-term (between five and 20 years), depending on the industry involved. Other organizations can only realistically look forward two to 12 years; their long-range planning is what other organizations would call medium-term planning. On the other hand, almost any organization should be able to put a credible effort into short-term operational planning, which is generally considered as one year.

Some organizations borrow from the military terminology to describe planning as strategic and tactical. They use the term *tactical planning* as a replacement for long-term operational planning. In many cases, this is a misnomer. The word *tactical* normally relates to smallscale actions intended to serve a larger purpose. This is not a normal business concept. Most businesses' plans are either routine operations, such as working on a systematic profit plan, or changes of direction, such as new strategies.

Long-range operational planning often occurs early in the year—before short-range and budget planning is attempted. The thought processes for short- and long-range planning are very different. Long-range planning is conceptual. It emphasizes management's long-term objectives over immediate goals that must be accomplished. In the MIS arena, this relates to the long-range operation of the business and its products. In this case, strategic planning is defined by the effort required to have an impact on a firm and its ability to compete through the development of computer systems.

Effective strategic planning can only be accomplished, of course, if the routine work of gathering data, comparing possible equipment and systems, and regular reporting of status has been done. This mundane work provides the foundation on which a strategic plan can be built.

Don't make the mistake, however, of thinking that all MIS planning should be labeled strategic. Most MIS projects are routine, not high-priority, and can be planned and handled by the MIS staff and related users with little input from top management. Such projects focus on operational improvement and ongoing maintenance, and usually can be handled within the annualized operational and budget planning cycles.

The aim of strategic planning for MIS may be summed up in two questions:

- How can the parent organization be modified to adapt to a new situation?
- How can the new technologies contribute to the overall success of the enterprise?

The goal, then, for strategic planning in MIS is to direct itself towards the concerns of top management. The purpose of strategic planning is to facilitate and support company management in making better decisions to improve profitability and meet company objectives. It is focused more on business strategy than on the systems strategy.

Strategic planning is not only for large companies; it works for small businesses as well. Contrary to what you might expect, you do not have to employ sophisticated planning staffs and computerized decision-making models to do strategic planning. In fact, small MIS Departments need in-depth, integrated, anticipatory thinking as much as do very large MIS organizations—if not more.

Technological and Personnel Performance Planning

Technology forecasting is often the responsibility of the MIS planning group, which creates a statement of future technological expectations and offers possible scenarios of specific technological opportunities in light of technological changes. This forecasting should be done whether the organization is forecasting strategic changes or will conduct business as usual.

Most companies rely heavily for their technology forecasting on the information received from their principal hardware or software vendor, listening to prepared advertising presentations and making their plans accordingly. This strategy works well enough for many companies. The problem, however, is that most vendors have far more presentations than can be attended by a particular individual. How do you choose which ones to attend? This decision can only be done realistically by a professional who has routine responsibility for, and interest in, the various concepts that are being publicized. This might present a problem when it is not possible to have a person doing technology forecasting regularly. The most practical solution is to form a group of companies with similar problems for semiannual discussions of the systems that membership organizations have developed.

Personnel performance planning, another important operational planning area, might not be the exclusive domain of MIS planners, but is of interest to them. Personnel performance planning is the basic responsibility of the management of each operational unit, under the general approval and direction of the personnel department. The output from such planning, however, is key data in operational planning. Therefore, the MIS planning group might be called upon to help set up a personnel planning system. A planning system develops a structure to upgrade hardware and software systems and consolidate data files.

The principal contents of the plan are *long-range action items*, including probable priorities and cost estimates. The plan should outline proposed projects and their goals as well as suggested times for management to review the systems activities. The plan should also address personnel issues, defining the proposed responsibility within the organization for executing the planned activities, and providing a schedule. A common problem with long-range planning is that it is time-consuming. Relatively few managers are able to shift their focus from controlling the bottom line to developing a long-range plan. Often, senior technical people, with little cooperation from managers, end up doing the planning. The existence of a long-range plan, however, is considered a necessity by most senior managers. It ends up as part of their annual plan.

In the end, the MIS department produces a document that includes an evaluation of external environmental forecasts, a review of organizational management considerations and stated objectives, a brief history of existing systems, and a vision of the future. The statement shows estimates of all projected costs, including training costs, and outlines the most likely requests for management.

Project Planning

Project planning is central to all system development and system installation efforts. It is normally handled completely within the systems development and operations departments, with no MIS input. The planning group, however, must receive documents throughout this process. In many cases, the project planning group integrates the automated system for project control with other reports. These other reports include regular plan-vs.-actual reports and capacity planning reports. Input also comes in the form of system specifications, system and program documentation, operations manuals, and user reference manuals.

Financial Planning

Of all the planning categories, financial planning is the most highly structured and clearly defined. Financial plans are produced after senior management goals and assumptions have been communicated along with a request from financial management for more detailed instructions.

The financial plan includes an annual expense budget that covers cost distribution as well as any financial projections for a single year. It also includes a *capital budget plan* that covers fixed assets such as land, capital equipment, and buildings for a year.

A financial plan might also include a departmental profit plan and a staff level plan. Departmental profit plans are set up for "soft money" transfers of funds within organizations.

Some organizations want to see certain departments as "profit centers." Financial planning for such organizations should take into account this strategy. On the other hand, a staff-level plan focuses solely on staffing needs and costs. Departments with strong personnel needs—often half of a departmental budget—frequently use this procedure.

Chapter

10

The Buying Team

Your team should be able to find many reliable sources of information on imaging systems. The problem is how to select the best sources. Although you can rely on vendors for education about imaging technology, you can also get enough information on imaging technology from periodicals, conferences, and trade shows that you can ignore vendors altogether. The best strategy, however, is to get information from every source you can, including vendors. The first step is to educate your team, making sure they are image-technology literate, as well as focused on reaching specific organizational, technological, and project goals before you start spending time with vendors.

A good starting point in the educational process is contact with the Association for Image and Information Management. As a trade organization, AIIM offers members many benefits, including access to an extensive resource center that offers numerous reports and publications on imaging. They also have a specialty bookstore where you can buy imaging technology reports as well as books on subjects ranging from proposal writing to scanner technology. The organization also sponsors the largest annual trade show for the imaging industry. (For a list of some of the reports, publications, conferences, and trade shows of use to your organization, see appendix A at the end of this book.)

You might want to start the education process with a broad understanding of electronic imaging technology. Then you can quickly move onto collecting more detailed and project-specific information. Vendors can often be helpful in this phase. Many will provide an imaging education to the prospective buyer during the requirements analysis and systems design phase. The payoff for the vendor is that they get more information about the buyer's needs, so this education process is a two-way street.

In the end, you'll want to keep the educational process to a minimum. The objective in bringing imaging into your organization is to simplify your life, not make it more complex. No one expects you to earn a college degree in electronic imaging. Of course, just studying this book will provide you with much of the electronic imaging information you will need.

Image System Requirement Analysis

In the requirement analysis phase, keep in mind that you can never ask every question you'd like to ask. As questionnaires get longer, survey participants tend to get tired and less likely to answer your questions. Focus your questionnaire to get the most important information from participants.

Questionnaires

One way to keep a questionnaire focused is to set objectives. These objectives might be as simple as, "We want participants to gives us a list of reasons they like or dislike our proposed electronic imaging system." Also, set priorities. Questions addressing high-priority issues should be included; questions addressing low-priority issues can be discarded.

Creating a questionnaire helps give your interviewing work a structure. The best questionnaires include both closed and open-ended questions. Here is a typical closed question:

Once you have a document, how long do you usually use it?
__ Less than one minute
__ One to five minutes
__ Five to 30 minutes
__ More than 30 minutes

This question allows only one answer, making it easy to analyze and tabulate results. Because they don't take long to answer, you can ask many closed questions. On the other hand, you only get the information you ask for, nothing more.

These are typical open-ended questions:

- What documents do you use most often?
- Do you think our organization needs an electronic imaging system? Why? Why not?

The disadvantage of using open-ended questions is that they often result in more unfinished surveys. People just don't want to bother answering them. This can add administrative costs to your project as you need to contact more people. Also, interpreting the result of open-ended questions requires extra analytical efforts. On the other hand, longer, more involved questions can provide information that shorter questions cannot.

Probe questions are open-ended questions that add depth. Typical probe questions start with, "Is there anything else you want to say about . . ." or "Are there any other reasons you can think of . . ."

Closed questions can be made open by the use of follow-up questions. For example, you could ask the closed question, "Do you like the idea of using hand-held scanners in your workplace?" The only two real choices are *yes* or *no*. However, if you follow up with the question, "What do you like or dislike about hand-held scanners?" you end up with the best of both worlds. You can tabulate the closed questions immediately. Later on you can tabulate the follow-up responses.

Both open and closed questions should be carefully worded. Here are a few suggestions:

- Avoid jargon. This is especially important if you are technically-oriented. Remember that the persons you are talking with might not "speak your language." Speak theirs.
- Keep the tone conversational; otherwise, you can end up with stilted questions that only confuse everyone.
- Don't talk down to people. You can be too folksy; talking like this will sound phony and unprofessional.
- Avoid loaded and leading questions. Ignore the temptation to design your questions to get the answers you want. As much as possible, don't put thoughts into people's heads.

A classic example of a loaded question is, "Are you still beating your wife?" In the imaging arena, a loaded question would be something like, "Do you think we should save our organization by creating an electronic imaging system?" Try saying *no* to that one.

A leading question, on the other hand, is a little more subtle, but also forces results a certain way. A political example illustrates the point. A poll conducted in May 1977 asked the question, "Do you favor the United States continuing it ownership and control of the Panama Canal at the end of this century, or do you favor turning the Panama Canal over to the Republic of Panama?" Eight percent approved the treaty; 78 percent disapproved. On the other hand, an August poll put the question this way, "The proposed new treaty between the U.S. and Panama calls for the U.S. to turn over ownership of the canal to Panama at the end of the century. However, the U.S. will maintain control over the land and installations necessary to operate and defend the canal. Do you approve or disapprove of the proposed new treaty?" This time, 39 percent approved of the treaty and 46 percent disapproved.

Interviews

Just because you have completed questionnaires doesn't mean you are ready to conduct interviews. You need to pretest the interview, talking to five or ten people to find out if you have any questions that are confusing or unanswerable. One of the occupational hazards of interviewing research is that you can get so immersed that you can start taking for granted the notion that everyone thinks just like you. Of course, they don't.

The interview process is not only an opportunity for you to collect information and to solicit input, it's also an important chance to educate constituents about the importance of an imaging project. Make sure that your interviewers are well-informed and enthusiastic. Although the interviewing process is an opportunity to proselytize the cause, the interview should also incorporate input. For example, you might want to solicit ideas or questions about current and proposed workflow. You should also be aware of the emotional impact of proposed changes and not be afraid to solicit employees' thoughts and feelings about imaging in the organization.

Sloppy interviewing produces sloppy results. The best way to prevent errors is to make sure that interviewers understand the importance of their work. At minimum, they should know the difference between a good interview and a poor one.

Questions should focus on document flow. You might want to diagram how each document is used within each job function (although some details, such as image frequency, cannot be incorporated into a diagram). Your questions should make this diagram possible. Some typical objectives to build into the questions include

- Estimating the number of copies of documents created, used, and stored
- Detailing the origin of documents and their destination
- Estimating usage frequency
- Identifying the number of documents that are created, used and stored
- Detailing hardcopy and screen requirements for documents

You will also want to use the interviewing process to piece together first a detailed description of the working environment and then an analysis. This is a chance to inventory all of the aspects of existing systems. You might want to start with an inventory of job functions. Questions should identify computer aptitude and experience of each job function or establish the scope of tasks involved. Another objective of the interview is to detail the amount of time spent locating, retrieving, copying, and replacing documents.

One of the most important things the interview process can accomplish is to highlight operations that are currently deficient, wasting time and resources. Be on the lookout for bottlenecks or "black holes" in the document flow, and deficiencies or variances in procedures. These findings should be formally consolidated into a summary of deficiencies and anomalies. These findings should improve methods of operation regardless of whether new technology is added.

The interview process can give the leader of the imaging evaluation an opportunity to educate others about imaging technology and its benefits, and introduce them to those with interest in and enthusiasm for the project. These individuals can offer a broad and well-informed perspective as well as form the nucleus for the team that will conduct evaluations. Politically, this is a time to show that the project has no preestablished agenda, that you and your team are interested in everyone's ideas and opinions.

Presenting the information

If you don't publish a final report for the interview phase, you should at least consider presenting a series of diagrams with supporting text. The diagrams should depict the current flow of documents through the organization and identify each work process performed on the documents. Consider using a limited-function CASE tool to assist in the preparation of the diagrams if the imaging project is examining more than a small section of your organization's operations. In any event, this information should be supported by a series of text descriptions detailing the current document flow.

These charts will help point out the true advantages of electronic imaging and help you focus beyond the easy-to-reach applications and toward redesign of your organization's most critical operational functions. Your imaging system should provide an online measurement of the performance of those processes that are integrated. The importance of online performance monitoring cannot be overemphasized. Without it, organizations must rely on reports of past performance to guide present and future activities; something like steering a boat by watching the wake.

Meeting Checklist

Coordinating your buying team's activities involves the coordination of meetings. Obviously, you will want to keep these meetings as short and effective as possible. In order to maximize the efficiency of these (or any other) meetings, consider using the following checklist to break down the keys to meeting success.

Notification

- Letters
- Media contact
- Bulletin board
- Personal contacts

Agenda

- Copies of agenda
- Material needed
- Committee reports
- Minutes

Location

- Room
- Transportation
- Electrical outlets
- Acoustics
- Room size and shape
- Parking facilities

Equipment

- Tables
- Chairs
- Microphones
- Tape recorders
- Video recorders
- Slide projectors
- Camera and film
- Extension cords
- Projectors
- Computer terminals
- Typewriters
- Bulletin boards
- Easel
- Duplication equipment
- Broadcast equipment

Materials and supplies for meeting

- Chalk
- Pens and pencils
- Paper
- File folders

- Computer disks
- Staplers
- Masking tape
- Copies of minutes
- Copies of reports
- Books
- Training manuals

Budget

- Rental
- Mailings
- Telephone
- Paper materials
- Secretarial time
- Transportation
- Hospitality
- Speaker and consultant fees
- Tape and film reproduction
- Entertainment
- Other

Functional Specifications

Functional specifications usually describe how applications interrelate within a system. However, if the application represents more than one system, the functional specification describes the relationship between systems. Either way, the functional specification should provide enough information to allow you to summarize your requirements across the applications. For example, if your system includes six applications, each of which requires three databases of images, and each database needs to accommodate up to 10,000 images at a time, you can calculate a requirement for storage of 180,000 images.

You must identify all points of linkage between the proposed system and your organization's existing information systems. You also need to collect detailed information on how hardware and software will function for these revised workflow diagrams. While these specifications should include hardware requirements, they should not include detailed software or hardware configurations.

How long should the functional specification document be? If you only have a few applications, it could be as brief as 20 pages. If the applications are enterprise-wide, involving major operational restructuring, the document could run hundreds of pages long.

A detailed description or analysis of the system design for an imaging system is a comprehensive, lengthy document. Each organization's is unique, based on its application mix and many other factors. The functional specification should include such details as the following:

- Image traffic volumes
- Database indexes and search schemes
- Document capture, retirement, and purging schedules and volumes
- Document image display and hardcopy requirements
- Workflow automation requirements

Once your functional specification is developed, you can summarize your organization's requirements. After that, you can conclude your team's education in imaging technology, and begin to talk with vendors. At that point, start thinking seriously about costs.

Project Cost Estimates

A well-prepared functional specification is a first step toward getting the vendors' serious attentions. It is also a valuable tool for your organization. With a specific understanding of your organization's needs, you can scrutinize the vendors, find out more about products, and start to develop approximate costs for your project. Most vendors have, in slang terms, "sweet spots" in their product mix. In other words, each vendor has a best-valued configuration that supports some, but probably not all, of your applications. When you meet with different vendors, note which ones come closest to your needs or show promise. This is the point where your team completes its background work by surveying currently available products. Vendors will be anxious to help and usually can offer considerable technical and management consulting.

Vendors are aware of how long the purchase cycle for most imaging projects is. They are most inclined to focus on projects that have already been granted management approval. This does not mean that your project must have a budget or final approval, but if you insist on talking to a vendor before preliminary approval of the project is granted, you are likely to be turned over to less-qualified staff.

The vendors should be able to help you develop cost estimates based on your functional specification. The product and pricing information they provide, however, need not be binding. After all, the intention is not to gather information to be used to select a vendor, but rather to identify a pool of vendors from which a final selection can be made.

Use a common matrix to tabulate each vendor's costs. Be sure to include both recurring and non-recurring costs. Non-recurring costs usually occur in the first or second year, but it depends on how your installation is staged. Break recurring costs down by type and year. The ultimate goal is to come to an analysis of yearly costs.

After you have narrowed your choice of vendors and solicited performance and cost information from them, you will be able to establish a range of approximate costs for your project—often a wide range. Remember that costs are never the only factor guiding your decision. Subjective factors always come into play. Your team is likely to select a vendor who inspires confidence and avoid those who leave you uncertain or uncomfortable.

Current costs determination

Costs can be broken down into expenses associated with

- Equipment
- Personnel, including salaries and benefits for all levels of management, advisory, technical, and support staff
- Equipment, including both high-tech systems, such as workstations and software development, and low-tech systems, such as file cabinets
- Maintenance (Never forget that it costs money to run equipment and keep it running, as well as to buy it.)
- Supplies, covering the cost of filing, copier supplies, and microfilm
- Overhead costs, such as insurance, security, and lighting

Compare these approximate costs with known expenses of the current methods of operating. The current costs might be readily available from the organization's accounting system, but it's more often necessary to estimate the costs of operations.

Ultimately, you should evaluate a system in terms of *opportunity costs*. More of an economist's term than an accountant's term, an opportunity cost represents the price of not pursuing an alternative. Inefficiency is an important opportunity cost associated with a decision not to use electronic imaging. For example, two important opportunity costs of keeping a low-tech filing cabinet system are searching for misfiled documents and recreating missing documents.

The highest opportunity costs are associated with lost revenues due to delays in cash flow. Electronic imaging, therefore, is exceptionally valuable in the banking and insurance industries where delays in internal processing of accounts can be costly. Suppose, for example, that current procedures in a bank delay the posting of accounts for five days, and that an electronic imaging system will reduce float time to one day. In effect, the opportunity cost of not converting to an electronic imaging system would be the costs associated with slowing down the workflow.

Suppose the interest rate is five percent per year, and that the bank handles $100,000,000 worth of deposits. Delays in deposit posting mean a delay in being able to lend money out again for an entire week. Five percent divided by 52 weeks, multiplied by $100 million, equals a loss $96,150 per week. Under the new system, the bank would cut this loss by eliminating ⅘ of that float, reducing the cost of the amount of money unavailable to the bank by $76,920. The opportunity cost for using the old system would therefore be $76,920 each week.

Costs, benefits, and risks

Sometimes, however, traditional costing methods underestimate the true benefits of a new system. *Cost/benefit analyses* are used to evaluate a current procedure that is being revised to result in a new procedure. You don't need cost/benefits analysis to evaluate an imaging system that can best be described as records management with little workflow process.

Cost/benefit analysis starts with a two-step process. First, the costs of the old procedure are compared to those of the new procedure. Then the benefits of the old are compared to the benefits of the new. The final result is an anticipated savings (or loss).

Like any investment, the implementation of an imaging system involves risks. The decision to introduce a new system requires an understanding of the nature of the risks and, if applicable, a comparison of these risks with those incurred under the present system. Too many organizations avoid quantifying risks because their financial practices deal only with "tangible" elements, and risks are often difficult to quantify. However, all potential risks need to be identified and, if possible, estimated in terms of costs before a decision is made. If nothing else, an analysis of risks is needed to estimate replacement costs for the system. Typical risks for a proposed system would include the following:

- The use of technologies unproven for the application
- Dependence on personnel who might not be available
- System failure

Risk evaluation tends to be easy and straightforward when there are few procedural revisions. However, in applications where workflow is being restructured, risk evaluation can be difficult. Often an organization's accounting system is not able to provide a basis for concise cost projection. Projecting the value of benefits involves uncertainties. Still, cost/benefit and risk analysis are useful in circumventing traditional accounting's tendency to underestimate the true value of electronic imaging. However, even these more advanced tools can underestimate the true worth of electronic imaging. An AIIM resource report entitled "Feasibility Studies for Document Management Systems" sheds light on some limitations organizations might impose on this phase of analysis:

> "Organizational members may perceive problems and opportunities differently, making problem identification a source of controversy rather than consensus. The same limitation applies to the specification of objectives and to the specification of alternative systems. Furthermore, there is no guarantee that measurement and analysis of alternative system costs and benefits will occur on an even basis. Not all costs and benefits are quantifiable, and those that are may not share the same degrees of concreteness, relevance, and so on . . .. Finally, outright 'political' posturing and advocacy of favored solutions may bias study efforts."

Ultimately, risk is a criterion for judging the relative value of different options. For example, you might want to select a strategy that offers maximum gains and yet risks the possibility of total failure. Or you might decide on a less risky system that will produce good, but less-spectacular results. There are always trade-offs. Ultimately, risks have their own cost/benefit ratios.

For example, many imaging systems today use OCR devices for scanning in data. OCR, however, is not risk free; data conversion projects demand a high degree of accuracy. OCR usually can operate with error rates of less than ten percent, but if the controls are difficult to implement, error rates can climb. Keyboarding, on the other hand, is a much more proven technique. Conservative systems designers sometimes prefer this less risky method. Keyboarding is slower, though. The trade-off, then, is between speed and accuracy. In some cases, a hybrid system can be set up to take advantage of both technologies. However, even if some keyboarding must be used to correct critical entry errors, OCR might still be the cost-effective technique.

Quantitative models can highlight these differences. Let's take, for example, an inventory application in which the spelling of the names of parts might not need to be exact, as long as part numbers and quantities are accurate. For the sake of illustration, assume an inventory of 20,000 items, with critical information occupying 10 percent of a 1,000-character record identifying each item. The remaining 90 percent is devoted to descriptive information.

In this example, the total number of characters to be captured is 10 million. At a 90-percent accuracy rate, the gross number of corrections will be 2,000,000 characters. However, if only 10 percent of these need to be accurate, the number of critical characters will be 200,000 characters. Will such a system be cost effective? Yes, as long as the cost per character for the OCR machine and whatever adaptations have to be used is less than the cost of keyboarding critical information. In this case, the cost of correcting 200,000 characters would have to be less than the difference between the cost of typing 2,000,000 characters by hand and the cost of inputting 2,000,000 characters via a scanner.

Cost planning for operator specialization

Planning personnel costs for imaging systems is tricky. Not all systems tasks require the same degree of complexity. There is a natural temptation to foist the more difficult tasks on those operators who exhibit the talent, aptitude, and willingness to take on difficult tasks. The upside of this strategy is that complex tasks are likely to be accomplished faster if performed by the operators perceived to be superior.

Some consultants recommend an array of salary schedules that permits average salaries to be lower if many of the operators perform only routine, easily teamable tasks for lower wages. This strategy is likely to cause perception and motivation problems. Everyone is sure to observe that some operators are more highly regarded (and paid) than others. Those in the lower categories will hold the usual mixture of resentments and aspirations as they view that gulf. High-tech employees, moreover, tend to be less amenable to these differences than people in other fields.

The only solution is careful planning. You must carefully plan the degree of specialization you plan to incorporate before you do it. It might be wise to feel your way through your own application before you publicly commit to pay and skill differences among technical operators.

Improvements Through Upgrades

In many cases, profitable improvements can often be made through system upgrades of key hardware and software. Such improvements require significant financial outlays for each item added; each should be justified separately. Common areas for system upgrades include the following:

- Electronic memory and storage
- Online peripheral devices
- Offline peripheral devices
- Intersystem compatibility

Memory upgrades can be accomplished by adding RAM to the CPU or by adding built-in peripherals with their own RAM. Changing offline disks, tapes, and disk storage is another important way of adding memory. The primary reason for adding memory is that a system's volume and nature has changed. Sometimes, however, addition memory is needed because new, desirable software demands it.

Both RAM and disk storage are expensive, although less so than they used to be. You can cost-justify this type of expansion in one of two key ways:

- For labor savings resulting from less frequent movement of data (by the operator) from one form of electronic storage to another
- For the ability to hold and process new software requiring additional RAM that might have previously been unavailable

Upgrading online devices or peripherals, such as terminals and printers, is often necessitated by new workloads, technological improvements, or cost reductions. For example, a few years ago, you could not find a laser printer for less than $2,000. Now low-end laser printers regularly sell for less than $600. Users need to display resolutions of 1,024×768 pixels and 256 simultaneous colors to comfortably view grayscale and color images on their computer. A 600 dpi laser printer with six megabytes of page memory is the minimum setup capable of printing newspaper-quality grayscale photographs.

Many consultants recommend that as your needs and perceptions change with experience in system operation, you should automatically think about profitable additions or replacements for peripherals. Because the technology moves fast enough to make printers obsolete in a few years, savings available from a new one can actually pay for "premature" replacement. Many users keep their old devices and simply add the new-design models as their need rises. Thus, they can exploit their original investment while keeping abreast of technology. Of course, a system needs to have sufficient hardcopy input to justify scanner costs.

Networks are justified by the operational advantages of linking new systems. Intersystem compatibility can become an issue any time a new system has to interface with an old one. In the age of enterprise computing, this is increasingly common. In most cases, such problems are solved by the use of software products. Information from foreign sources received on electronic media can be converted by devices or software that translate among the various word processing protocols and retain embedded codes.

The irony is that "almost-compatible" components cause more headaches than any other factor in incorporating low-cost add-ons. Any trace of incompatibility usually causes enough trouble in routine operations to seriously threaten any hoped-for advantages. Systems owners compensate by purchasing or inventing hardware and software solutions.

The generic disk unit, for example, is a frequent problem area, especially when you select one that seems extremely attractive compared with the usually expensive equivalent from your original system supplier. It might, in fact, be useful. Interfaces with disk units, however, are often touchy, and "only slightly different" equipment can turn out to be highly incompatible with your system. Although the third-party disk manufacturers do assist with installations, they do not consider themselves responsible for total compatibility with every existing system. Unfortunately, sometimes third-party disk units end up being discarded.

To avoid problems, the best precaution is to require full demonstration of all equipment purchased. Sometimes, however, this strategy is not feasible. For example, vendors will be reluctant to let you test small items or pieces that have to be shipped great distances. The next best step is to create contractual language in the purchase request, requiring that payment wait until satisfactory performance is observed. This effectively shifts the burden of compatibility on the supplier.

The Quality Issue

Whether buying a new system or upgrading an existing one, quality is always an issue. Consultants warn, however, that you should avoid solutions that attempt to trade off a little less quality for a lot less cost. Solutions that seem less expensive in the short-run can have long-run consequences. Issues to be investigated include:

- *Reliability.* How long will the item stand up? If it breaks down, what is the cost of fixing it? Indeed, can it be fixed at all?
- *Replaceability.* Will the supplier still be doing business if you later need to replace the add-on?
- *Auxiliary expenses.* Must you modify any existing portions of your system just to integrate the new component?

Imaging Systems as Investment Opportunities

Whatever system of evaluation you use, in a profit-making organization, any expenditure must result, directly or indirectly, in the production of income. In a nonprofit organization, any investment should demonstrate an improvement in the use or allocation of resources. Either way, an automation project is primarily viewed as an investment opportunity rather than as an expenditure. The project can then compete with other business opportunities included in an available portfolio of investments.

Financial justification usually relies on analysis of two of the most widely used financial indicators:

- Return on investment (ROI)
- Payback period

In its simplest form, an ROI calculation takes place in two steps. First, the net cost of a project is compared to after-tax savings, then these savings are translated into the rate of investment that would generate the same amount of interest for the same expense.

One of the key drawbacks with ROI planning is that savings projections assume that business conditions will remain stable. That is why many companies rely on payback periods. The concept is very simple: the quicker a project can pay for itself, the more desirable it is. As a rule, given two projects with equal ROI, managers generally support the one with the shortest payback. In the absence of an emergency so severe that imaging is mandatory, the payback period should be two to three years. Sometimes, unexpected cost savings can arise. For example, a major savings institution once installed an imaging system to handle loan applications. Those applications contained in-depth information about the applicants. The marketing department for the same institution, meanwhile, was buying mailing lists to solicit business. Sometimes they bought the same names as those whose loan applications were being processed. The marketing department ended up sharing information from the loan department to build a marketing database.

Most enterprises today don't have to look far to find potential benefits. Often the problem is deciding which issues are most pressing. Quantifying the opportunity costs of these problems in several different departments is often the most efficient way to decide where to start. For example, the New Orleans police department cited reduced collation mistakes and improved data integrity as justifications for purchasing an electronic imaging system. McDonnell Douglas justified expenditures by pointing toward a reduction in personnel costs for mail handlers, file clerks, and data entry operators.

Unfortunately, many of the benefits electronic imaging offers are harder to quantify. For example, conversion to an electronic imaging system might reduce legal exposure resulting from lost documents. However, just how much money will be saved is hard to say. Certainly, the last thing anybody would want would be to test the hypothesis in the real world. Computer modeling, however, can help by running "what if" scenarios. Possible examples of intangible benefits include:

- Better backup
- Improved security
- Opportunities for new products and services

Chapter

11

The RFP

There is no fundamental difference between preparing and answering an RFP (request for proposal) for an imaging system and for any other computer system. Appendix D shows an actual imaging RFP, in case you have not seen one before. Also, the sidebar at the end of this chapter details the preparation of a LAN imaging RFP. The rest of the chapter is intended as a short course on RFPs in general.

An RFP accomplishes the following:

- Documents all areas of requirements
- Fixes the form and format of supplier response to those requirements
- Lays down a fixed schedule for receipt of supplier responses

The idea of an RFP is to establish a partnership with your vendor. People who have written many RFPs suggest that an adversarial posture be avoided, at least until a system has been fully specified. The time for confrontation is when all issues besides contractual ones have been cleared.

One common mistake made by people writing their first RFP is to rush the process. You might think that you can rush through the presentation of the RFP and then make changes during the final negotiations. However, the reality is that there will be plenty of other things to change by then. Miscalculating or misstating needs early on will only mean more wasted time for you and your vendors.

The RFP needs to be evaluated internally before it is presented externally. You can shorten the internal approval process by setting up informal review sessions with management and board members. In many cases, the next step involves preparing a formal presentation to the management or the board to address all questions and concerns. As part of the preparation process, you should, of course, know what these concerns are. In many cases, board members will ask for additional information or analyses. Informal reviews beforehand minimize that risk and are also a way of avoiding future conflicts. Sometimes vendors attempt to bypass the selection process by influencing the senior management of your organization directly. Your team's primary defense against this kind of end-run is a well-established relationship with your decision-makers.

If the RFP is extremely long and presents particularly tough problems, consider scheduling a conference before you actually issue it. This conference allows vendors to ask questions about your RFP and any aspects of your decision-making process. The conference also offers the RFP team a chance to calibrate the vendors' expectations, and then make any final changes. After these changes are made, you can formally release the RFP. In most cases, vendors will respond in four to sixteen weeks. Actual response time depends on the complexity and technical depth of the RFP as well as imposed deadlines.

The Team Approach

RFP preparation often requires input not only from writers and editors, including proposal specialists, but from technical staff as well. That's why many companies are turning to the team approach to RFP writing. These teams include a full range of engineers from several disciplines, including quality control, testing, program management, service and maintenance. Marketers also are team members, providing competitive analysis. Legal and accounting as well as personnel offices also lend their expertise to the process. In other words, in many companies, lawyers, accountants, and technical experts are not just contributors to RFPs. They are coauthors.

Each RFP, however, should have a single proposal manager. This manager should have the respect of the proposal team. Ideally, he or she should be the team member who is most excited about a project and provides the most energy and direction. His or her job is to be the ultimate authority on an issue, available to answer questions, make critical decisions, settle arguments, and keep a proposal on schedule.

Preparing an RFP

Because it is an important tool, great care needs to be taken in the preparation of an effective RFP. Preparing the RFP gives your organization an opportunity to establish system specifications. These specifications should detail the optimal imaging systems and its applications.

You might want to hire independent consultants to help with system specifications. As the old saying goes, "There is no substitute for experience." Experienced consultants can provide fresh ideas and accelerate the RFP process by using their accumulated expertise. Their job is to work with your staff to provide the technical detail needed for areas such as:

- Development
- Delivery
- Installation
- Maintenance
- Testing
- Training

The work together will progress in stages, starting with the preparation of specifications. The preparation of specifications is a rigorous process in its own right, discussed in chapter 10. Then, it continues through the detailed review of those specifications.

Help preparing an RFP

Vendors can help you prepare specifications if you invite them. Some RFPs contain clauses that call for vendors to collaborate in creating these specifications. Other times, your team can get this help before the RFPs are drafted. The specification process includes the detailing of hardware requirements, but goes beyond that. Integration specifications and systems support need to be covered. Workflow description revisions and workflow diagram revisions need to document the changes in the work environment. Training specifications need to be included. The physical limitations of your worksite require that facilities requirements be incorporated into the RFP.

A good place to start is to review successful RFPs from organizations similar to yours. You can use these successful proposals as a benchmark. In many cases, vendors themselves can direct you to customers who have faced similar problems and have developed RFPs that they are willing to share. The Association for Information and Image Management can also direct you to these sources.

Your legal department will want to review your RFPs periodically to identify problems and potential problems, and to prevent the possibilities of protest arising over a contract award. The legal department should check for the appearance of bias, which is considered the leading cause of contract protests. This is especially true for public-sector organizations.

Contract protests do happen in the electronic imaging field. For example, vendors successfully protested the awarding of a $1.4 billion contract to AT&T in the summer of 1991. The contract called for AT&T to provide the

Internal Revenue Service and other treasury agencies with 32,000 Unix-based microcomputers, 50,000 workstations, printers, networking equipment, software, and integration services. As a result of the protest, delivery of computers and other equipment was delayed for several months.

Sections of an RFP

Customize your RFP to reflect the unique needs of your organization. The RFP should embody your organization's operational goals and reflect its technological goals. It should make sense economically. Certainly, this type of information will make your RFP a lengthy document. However, failure to provide proper background information or misstating your needs will only increase the time you need to spend negotiating for your system.

Background information should describe why a project was initiated and briefly describe your existing MIS program. That way, vendors will know more about your organization's true needs. This background should also include information on how the electronic imaging system fits with existing systems. Also, summarize the specific applications you are seeking.

Next, provide the vendors with clear guidelines for preparing and submitting their proposals. When reviewing the proposal, you don't want to have to read between the lines. On the other hand, don't reveal the specific formulas or rules you will use to evaluate proposals. Your guidelines should include issues such as the following:

- Procurement timetable
- Subcontracting
- Reporting

One final procedural element is information about your organization. Be sure to tell the vendors about any unusual aspects of organization structure and policies. After all, the goal is to find a vendor's corporate culture that is compatible with your organization.

One convenient way to divide up the RFP team is to have one person in charge of each section of the report. A typical RFP will have the following sections:

- Executive summary
- Technical section
- Management section
- Pricing section

Because imaging is technologically oriented, the technical section will probably be developed first. The primary purpose of this section is to enable vendors to demonstrate how the products offered meet the RFP requirements. Vendors vary in their response to this section. One strategy is to make it very detailed to give the potential buyer all the information required to make an intelligent choice. In other words, the goal is to be as cooperative as possible. Another popular strategy is to keep details to a minimum because a failed bid could end up in the hands of a competitor.

The audience for the technical section, usually composed of technical people themselves, looks for an explanation of the technical aspects of the vendor's proposed solution, including product information and information on installation and implementation. Their level of technical expertise will be higher than for any other section. Still, avoid writing "over their heads." Any nonstandard or proprietary features should be explained in detail. On the other hand, the writer should not be so narrowly focused that he or she forgets the human element of this section. For example, readers of this section will want to know information about key personnel.

The management section builds upon the technical section. In many cases, the management section can be incorporated into the technical section. However, if the RFP calls for one, a management section can provide additional information, addressing questions about how a vendor will deliver the solution offered in the technical section. Personnel is emphasized, as are the physical resources required for the project. This is the best place to talk about needed training and project management policies.

The pricing section is targeted primarily toward accountants, but it will have a wide audience. Pricing is an artform and a science.

Although the executive summary is often placed first, it is usually written last. The importance of the executive summary cannot be overemphasized. Its primary audience is the high-level decision maker. Unfortunately, many of the highest-level decision makers have the weakest technical backgrounds. Because of this, one of the tasks of this section is to explain (and sell) the technical concepts in simplest terms, stressing the benefits of the proposed solutions. An executive summary should include the following:

- Recommendations for hardware, software, and integration
- Recommendations for training and support
- An assessment of organizational support
- A brief discussion of risks

Whenever possible, the vendor should take into account the competition when writing the executive summary. In other words, the idea is for the vendor not only to sell a proposal but to sell the advantages over the competition's solution, features, and benefits. As the old saying goes, "First impressions last." The executive summary's position as the first section people read increases its prominence. Unfortunately, few RFPs provide guidance for the development of the executive summary. A typical set of instructions might be as follows:

> This section provides a summary of the proposed system and all major components. It should contain a general system description, major capabilities as they apply to future requirements, and discuss any areas of concern which need to be addressed. It should be prepared at an appropriate technical level for senior management review.

Before the RFP is sent out, it needs to be reviewed by all people involved in its preparation. End users, technical and operations staff, and management all need to sign off.

Tips on editing RFPs and proposals

Revision is the real work of writing. In other words, writing is rewriting. Only very naive people believe that good writers do not need to revise, that letters, memos, manuals, and proposals simply pour forth in their complete and final forms. Any writer who sticks to this delusion will have little chance of improving his or her work. The best writers, in fact, are rewrite junkies. Ernest Hemingway rewrote the last chapter of *A Farewell to Arms* 17 times, and revised the last page 39 times. A good rule of thumb is to edit your work at least three times.

First-round editing emphasizes clarity and completeness. Ask yourself, "Have I provided all the information that my reader needs?" This is also the time to get rid of any embarrassing errors and start making cuts. The goal is to cut material without interfering with the reader's understanding.

In the next round of editing, think about your audience as well as the structure of the presentation. Writing to an audience is an art in its own right, based upon an awareness of their expertise and position. You do not want to end up appearing condescending, nor do you want to talk over people's heads. Knowing who you are talking to ultimately determines how you should discuss an issue.

This is also the phase to look at the mechanics of your writing. Remember that each paragraph should have a topic sentence that somehow summarizes the contents of the entire paragraph. Does each of your paragraphs live up to this basic premise? Indeed, one of the best ways to construct an outline for your work is to list the topic sentences for all your paragraphs.

The final round of rewriting is polishing. Remove any of the spelling, punctuation, or typographical errors that were overlooked earlier. Are your word choices exactly correct? Never assume that anyone knows what you mean when you write; the words must speak for themselves. Also, look at your sentences. Have you varied the lengths and the patterns to engage the reader's interest? How readable is your work? Finally, examine the tone, eliminating anything that would be seen as hasty, hurting your credibility, or offensive.

At the most basic level, an edited work should be devoid of mechanical, spelling, and punctuation errors. Above that, the document should be readable and well-organized. The presentation should be clear and concise. You want you readers to be able to understand you throughout your thorough presentation. Finally, you want to make sure that the document meets your audience's needs.

The RFI Alternative

Not every organization, however, uses RFPs for all purchasing decisions. A more informal way of getting information is through a *request for information (RFI)*. The RFI is a less arduous path for soliciting information and proposals. The issuing organization avoids the hassle of preparing a detailed specification. In turn, the vendors have an easier time answering RFIs because they do not have to tailor their discussion and proposal to the RFI's specification. RFIs make sense under the following circumstances:

- You have few bidders.
- The dollar amounts are small.
- Your organization lacks the personnel needed to prepare and review RFPs.
- The problem is minor.
- Your organization has addressed similar problems in the past.

One way to answer an RFI is with a *letter proposal* addressing the client's needs or problem and offering a solution all within the space of three or four pages. However, because a letter proposal is short, vendors have to be exceptionally careful about wasting space. A common mistake is to tell prospective clients what type of business they are in. They already know that. Nor do vendors need to reiterate business objectives. They are better off focusing on addressing specific organizational needs.

Whenever possible, vendors should quantify the benefits of the solution they are offering. Many managers tend to think in terms of numbers. This will also help you avoid the trap of confusing the benefits of a product, which are important to the vendor, and the benefits of solving your business problem, which is important to you.

The letter should conclude with a specific and clear recommendation to take action, namely the adoption of the vendor's solution. Be sure a vendor is as specific as possible about technical issues, including possible "red flags." After all, one of the purposes of the RFP process is to solicit technical information. The vendor should have already discussed alternative solutions. That strategy is a way of showing off knowledge of the business environment, and helps counter the competitor's strategy. An implementation schedule should also be included in the letter proposal.

The Final Steps

Once the RFP or RFI has been formally approved internally, you will want the attention of as broad a selection of vendors as practical. The RFP manager should prepare a list of suggested vendors and initiate formal or informal contacts with the vendor community to alert them to your intention to issue an RFP.

Most proposals are not awarded right away. The gulf between proposal acceptance and final award is wide and swallows up many worthy candidates. This phase requires selling, positioning, and negotiating; in short, constant interaction between potential vendor and potential customer. During that time, vendors will be able to build awareness of whether their proposal will be accepted.

After a vendor has submitted a proposal, it's up to the potential customer to act. Hopefully, they will ask for some follow-up. After receiving a proposal, you may ask a vendor for an oral presentation or question-and-answer sessions. A vendor might spend more than $100,000 for equipment and personnel to put on such a "dog and pony show." Any vendor should look forward to an invitation for an oral presentation. It does not mean that the proposal is somehow incomplete and needs attention. Quite the opposite is true, in fact. Being asked to make an oral presentation is an indication of serious consideration. In fact, in most cases, it means you've made the cut.

These questions have a profound effect on both the buyers and the sellers. For the buyers, they are yet another screening device. Failure to get clear, understandable answers will lead to a candidate disqualification. Sellers can use questions as a form of feedback to learn more about a customer's true needs. A good presentation will require additional resources, including time and money. The customer expects the presenter to have a thorough knowledge of all aspects of the proposal, including its weak points.

Sometimes customers forgo or postpone oral presentations in order to emphasize question and answer sessions. Many potential customers with a show-me attitude will also ask for product demonstrations. A demonstration is the last phase for dealing with technical aspects of a proposed solution.

Buyers, however, frequently have questions that go beyond the technical aspects of a project. For example, customers are likely to have questions about pricing. They will want to know how pricing policies were developed. If there are any nonstandard pricing decisions, they will want to know those too. The products selected themselves will come under scrutiny. Why were particular products chosen and others ignored? Finally, many vendors will want to peer into the decision-making process of the organization preparing the RFP itself, asking how the proposal was developed and who was responsible. In many cases, customers will be asked to make "what-if" predictions.

Questions from the customer are a positive indication to a vendor that its proposal is being given serious consideration. Vendors should give these questions as much attention as they give to original proposals. Most proposals have "soft spots" that require additional attention.

Meanwhile, research the vendor. Check references, visit or telephone the worksites of customers. The idea is to inspect a vendor's qualifications and equipment selection. Also, consider visiting the vendor's facilities to check for stability and the ability to deliver what is promised.

In some cases, technologies or other requirements change so fast that a customer might end the proposal review process with a "Best and Final Offer" stage. Public sector agencies, especially on the federal and state level, usually build this stage into the review process. This is the final chance for the potential vendor to make needed changes in the reply to the RFP. Any company that makes it to this stage should be confident that it is on the short list. However, just making it to this stage is not a guarantee of success. The best way to look at it is as the final hurdle before contract negotiations.

The final step before a contract is awarded is the contract negotiation. The purpose of this step is to get agreement for specific work statements, spelling out who is responsible for specific work and administrative tasks as well as a timetable. This stage is very much of a two-way street. Vendors and customers are encouraged to spend this time to build consensus. Not only is it important for both parties to understand what services and equipment are being purchased, but everyone must finish this step with a clear understanding of the limits of the vendor's responsibilities. Failure to reach this understanding will lead to constant problems once the contract goes into effect.

Reviewing an RFP

In most organizations, RFP opportunities are screened. Some salespeople do take a shotgun approach, responding to every RFP that comes across their desk in hopes that keeping their name in front of a customer will be a keen marketing advantage. They also reason that sales are a numbers game, and that "the more times you go to bat, the more likely you are to get a hit." In most cases, however, companies have screening committees.

Review components

One of the first steps in organizing a proposal is to have technical staff review the RFP. Their questions, in turn, shape the proposal-writing process. In a technical field such as electronic imaging, you cannot underestimate the importance of the technical staff review and its multitude of perspectives.

The individual components of a system require special attention. David Wood wrote in the June 1991 *Inform*, "Despite their importance, many display RFPs are inadequate today; display requirements are poorly understood and the technology is rapidly changing. Some RFP authors believe (mistakenly) that document imaging displays are commodity products which don't require specific attention in the RFP. Ideally, though, requirements for document imaging displays should be specified in such a way that users meet their requirements while vendors are given enough flexibility to provide the best price/performance ratio possible."

All document imaging systems, according to Wood, share three key display requirements that must be met for heavily used workstations:

- Screen size must be adequate for the application.
- Scanned images must be easily readable.
- Appropriate software drivers must be available.

In the display RFP example, each workstation in the system needs to be analyzed before writing the RFP to determine what those display requirements mean for that system. Among the key questions technical staff will ask are the following:

- What are the budget constraints?
- How many workstations will perform document imaging more than one hour per day?
- What are the sizes of the documents to be viewed?
- Besides the image, what else must be displayed simultaneously?
- What software will the system use?
- Are the images black and white, gray, or color?

The answers to these questions will translate into the display requirements of your system. Systems used more than one hour per day, according to Wood, should be classified as "heavy use." This is the most important factor, he argues. Heavy-use machines should be provided with high-quality displays that are clear and readable, and large enough to display a full page and the application simultaneously. Systems used less than that can be installed with an eye toward reduced costs.

The technical staff will of course, be asking technical questions. In the area of video monitors, for example, the proposal team will look at issues including the following:

- Adequate screen size
- Readability
- Video bandwidth
- Monitor spot size and brightness

Each of these areas demands a certain level of expertise. To understand technical analysis requires a certain level of vocabulary and a working knowledge of how imaging products work.

Screen size might not sound like an important issue, for example, but it is. In this case, the screen must be large enough to display the document along with related database management system information, as well as a 3270 emulation window. For $8^{1}/_{2}\times11$-inch documents, the screen size should be no less than 19 inches diagonal. A monitor less than 19 inches will create display problems, particularly the cutting off of valuable information. You can view smaller documents on smaller screens. A 15-inch screen is adequate for 5×7-inch invoices. Engineering drawings, on the other hand, would require even larger screens.

Readability has an impact on the overall efficiency of a system. The additional cost of a top-quality screen, then, has to be weighed against the possible loss from selecting less readable screens.

Coping with conflict

The temptation, of course, is to avoid all questions and conflicts. However, allowing conflicts to remain unquestioned only hurts the chances of a proposal's acceptance. The best way to view the questioning process is a positive experience for both the vendor and the prospective buyer. In fact, questioning should go on not only during the formal question-and-answer session, but throughout the presentation process.

Great care needs to be given to the phrasing and asking of questions. Some questions will be interpretative. For example, most imaging software today will run under Windows 3.1. However, some software engineers might prefer a more obscure system. Also, while most users prefer color over monochrome or gray displays, a case could be made for black and white display. This is especially true if budget is a consideration. Color displays cost three times as much as monochrome or gray ones. Other questions will arise from simple corrections of typos and other errors.

Copies of all questions and answers should be kept with other documents related to the proposal file. Of course, the proposal itself is likely to change as feedback is incorporated into it.

Examples of Technical Issues Facing an Imaging LAN

I. Software
- A. Operating system and utility programs
  - 1. DOS
    - a. Which version of DOS for the root directory
    - b. Multiple versions of DOS on the network
  - 2. Electronic mail
    - a. Menu-driven
    - b. Help screens
    - c. Display messages
    - d. Distribution lists possible
    - e. Notification when a message arrives
    - f. Ability to forward messagesg.
    - g. Ability to define multiple user groups
    - h. Ability to print and file messages

- i. Ability to include graphics
- j. Ability to attach files
- k. Other features desirable

3. Network calendar access to calendar features
   - a. Degree of workstations and hardware resources
   - b. Ability to schedule room

B. Network management

1. Ability to perform diagnostics
2. Ability to add and delete user groups
3. Password protection
4. Maintenance of user statistics
5. Ability to handle remote dial-in users
6. Ability to handle multiple operating systems
7. Ability to handle bridges to other networks
8. Ability to add and delete printers easily
9. Security provided
   - a. Log-in level (password)
   - b. Directory level
   - c. File level
   - d. File attributes for sophisticated users to bypass the menu
10. Menu-driven with ability
11. Log-in scripts or other factors making it easier for novice users to log in
12. Print-server software
    - a. Number of printers permitted on the network
    - b. Number of share devices
    - c. Printer queues
    - d. Commands available to network users for setting parameters for specific printing jobs
    - e. Commands available to network users for disabling printers for routine servicing
    - f. Print spooling software
    - g. Types of printers supported
      - i. Parallel
      - ii. Serial
      - iii. Laser
      - iv. Line printers
13. File-server software permitted
    - a. Size of volumes
    - b. Access speed
    - c. Number of logical drives permitted
    - d. Virtual drives
    - e. Restore-tape-to-disk capability
    - f. Directory hashing

14. Network communications server software
    a. Protocols supported
        i. ASCII asynchronous
        ii. 3270 BSC
        iii. 3270 SNA
        iv. X.25
    b. Speed supported
    c. Ability to handle callback modems
    d. Automatic dial-out
    e. User statistics provided

C. Current software that the network must support
1. Word processing
    a. Compatibility with current software
    b. Features required
    c. Training to be provided
2. Spreadsheet/financial analysis
    a. Compatibility with current software
    b. Features required
3. Database management
    a. Compatibility with current software
    b. Features required
    c. Training to be provided
4. Accounting
    a. Compatibility with current software
    b. Features required
    c. Training to be provided
5. Other application software required
    a. Compatibility with current software
    b. Features required
    c. Training to be provided
6. New application software required

II. Hardware

A. Microcomputers
1. Currently on hand (brand and configuration)
2. Additional workstations required for the network
    a. IBM compatibility to be able to run required programs
    b. RAM required
    c. Number of disk drives required
        i. No disk drives and an autoboot ROM chip
        ii. If disk drives are required, half-height or full height
        iii. 360K, 1.2M, 720K, or 1.44M disk drives
    d. Hard disk required
        i. Size (in megabytes)
        ii. Type of microcomputer

      - iii. Which version of DOS formatted for
    - e. Monitors and monitor adapter cards required
      - i. Color or monochrome
      - ii. Resolution
      - iii. Size
      - iv. Dual mode
      - v. Ability to run certain graphic programs
    - f. Other I/O cards required
      - i. Parallel or serial cards and cables
      - ii. Additional RAM cards
      - iii. Multifunction cards
      - iv. Accelerator cards
      - v. Others
- B. File servers
  1. Size required
  2. Processing speed
  3. System fault tolerance (if required)
  4. Other features required
  5. Number of tape backup units
- C. Bridges to other networks
  1. Other networks to be connected
  2. Adapter cards and cabling required
  3. Software required
- D. Backbone networks required to connect multiple bridges
  1. Description of bridges to be connected
  2. Processing speed required
- E. Gateways to minicomputer and mainframe environments
  1. Local or remote connections
  2. Protocols required
  3. Number of concurrent sessions required
  4. Terminal emulation required
  5. Local printer emulation required
  6. Amount of activity to be handled
- F. Minicomputers
  1. Currently on hand (brand and configuration)
  2. Need to integrate information with LAN
- G. Mainframe computer
  1. Currently on hand (brand and configuration)
  2. Need to integrate information with LAN
- H. Printers
  1. Currently on hand (brands, buffers, and accessories)
  2. Additional printers needed for the LAN
     - a. Speed required
     - b. Laser, dot matrix, or other

c. Compatibility with major printer drivers
d. Type of connection (parallel or serial)
e. Length of distance from workstations
f. Other special printer features
3. Special language fonts or downloadable fonts
4. Near letter-quality and fast dot matrix modes
5. Workstations or areas that will need to access specific printers
6. Unusual printing requirements (color, multiple copy, specific accounting forms, etc.)
7. Software packages on the network that require a specific printer

I. Modems
1. Currently on hand (brand, speed, special features, etc.)
2. Need for additional units with the LAN
3. Transmission mode required (simplex, half-duplex, or full-duplex)
4. Interconnections required (point-to-point or multiple drops)
5. Special features needed
a. Auto-dial
b. Auto-logon
c. Auto-answer
d. Other

J. Plotters
1. Presently on hand (brands and configuration)
2. Additional units required
a. Speed
b. Number of colors
c. Compatibility with major brands
d. Programs to drive plotters

K. Optical scanners
1. Currently on hand
2. Additional units required
a. Speed
b. Fonts to be scanned
c. Programs that will need access to this data

L. Other hardware required
1. Cash registers (for retail environment)
a. Type of connection (serial or parallel)
b. Compatibility with point-of-sale accounting programs
2. Badge readers (for manufacturing environment)
a. Employees clock in and out of several jobs for the same day
b. Information interfaced with an accounting program's payroll module
3. Multiplexers
a. Devices to be attached

        b. Location of these devices
        c. Type of transmission required
        d. Speed required
    4. Protocol converters
        a. Devices to be attached
        b. Protocols involved
            i. SNA
            ii. BSC
            iii. ASCII
    5. Power protection required
        a. Voltage regulation
        b. Elimination of sags and surges
        c. Prevention of common-mode noise

III. Vendor requirements
  A. Experience
    1. Company history (length at present location)
    2. Experience installing local area networks
    3. Customer references for similar installations
  B. Service
    1. Number of factory-trained service technicians
    2. Ability to provide on-site service
        a. Repair or replace service within 24 hours
        b. Ability to respond 24 hours a day
        c. Ability to respond within 2 hours
    3. A sufficient inventory of parts to provide adequate service
  C. Ability to provide a LAN demonstration
    1. Software to be identical with ordered software
    2. Hardware to be identical with ordered hardware
    3. Benchmark tests to be conducted
  D. Training
    1. Ability to provide basic user training
    2. Ability to provide training on all purchased software
    3. Degree of training provided with installation
    4. Charge for additional training
    5. Telephone support included in purchase price
    6. Kind of training for the network administrator
    7. Possibility of training others
  E. Responsibilities if multiple vendors are required for this network
    1. Hardware training and network familiarization
    2. Hardware service files
    3. Software training
    4. Software service

Chapter

12

Vendor Response Evaluation

After the RFP has been issued and vendors have had an opportunity to review it, schedule a vendors' conference. This conference provides an opportunity for unbiased contact between the vendors and your organization. It also lets vendors ask questions about the RFP, clearing up confusion and resolving any problems with the RFP.

Remember that at this point you are not choosing a proposal or vendor. Instead, you are selecting which vendors you will choose from. Eventually you will make a selection from this "short list" based on additional considerations. In part, you make the decision based on additional information the vendors submit after the vendor conference.

The formal evaluation of the proposals should be handled by a small team. Typically, you use a decision matrix. You might want to consider hiring an experienced, qualified consultant who can help set up the matrix and assist in its evaluation. In any case, the results demand a rigorous review conducted by your whole team. The process involves ranking products and assigning weights to various factors. You need to ensure sufficient review and discussion so that every member of the team supports and endorses the resulting rankings. Once the deliberations are over, your considerations and conclusions should be documented in writing.

Your review should focus on three areas:

- The organizational strengths of the vendors
- Costs
- The technical strengths of the vendors' solutions

Only you and your team can decide the relative importance of each of these factors. In many cases, your team can develop some formal, representative criteria in each area of evaluation. Many consultants recommend moving towards the elimination of subjectivity by developing criteria that are quantitatively measurable.

Whenever possible, you should develop formulas for measurement, preferably using tools that are derived from the RFP format. These tools can include the following:

- Spreadsheets
- Databases
- Paper-based evaluation forms

When properly designed and used, the tools provide an audit trail for the selection process, a trail that can be retraced to defend against a protest of the award.

Technical Review

The technical review of the vendor's proposal should use a matrix much like the one used to evaluate proposed applications internally. This is a four-step process for each application:

1. The evaluation process should focus on each application in the RFP, assigning it a weight.
2. Within the application, each element on the matrix should be ranked and each vendor's proposal assessed, then assigned a score.
3. The weighted score is computed by multiplying the rank of each element by the score.
4. The sum of the elements' weighted scores is the application's weighted score.

The final step is to find the proposal's weighted score. You do this by summing all the applications. This proposal weighted score then is the basis for quantitatively comparing and ranking the vendors. Like costs comparisons, performance comparisons may be tricky. A spreadsheet or modeling tool will remove most of the vagueness from this process, provide a clear and detailed picture of how the review took place, and show how each conclusion was reached.

You might want to use a standard evaluation form. Obviously, the evaluation criteria should not change from vendor to vendor. However, the factors for evaluating each application and their relative weights might change from application to application. The factors depend upon your organizational environment and your circumstances.

Cost evaluation should be based on proposals that are as comparable as possible. Your RFP lays the foundation for cost comparisons. In the real world, however, there are always slight differences between what vendors offer and organizations want. In other words, cost comparisons require some adjustments and guesswork.

Organizational Review

When conducting the review of the vendor's organization, pay particular attention to the area of field service. Surprisingly, unlike the computer industry at large, there are few if any third-party service organizations that support imaging systems, particularly scanners and optical storage devices. Carefully evaluate the vendor's service capabilities and confirm your evaluation with several current customers. You can use a pass-fail rating or a numerical, weighted score for this. Indeed, many consultants recommend this procedure for public entities subject to the challenge. Rankings based on service abilities also make sense when the technical proposals are evenly matched so that creating a short list is difficult.

Be sure, also, to look at the vendor's health as an organization. This means going beyond a look at corporate missions. Is the vendor financially able to meet your needs, or is their bottom-line health a matter of financial concern? Are they a growing company? Smaller organizations do not have the financial resources, by definition, that larger organizations do. This, however, can be compensated for by ties with larger, stable organizations. Small companies can also offer innovation, more support, and better responsiveness. The best way to find out about a vendor's abilities is to check their clients for a track record of similar collaborative relationships.

The Short List

The short list consists of those vendors whose proposals continue to merit consideration based on the weighted proposal scores that the cost, organizational, and technical evaluations provide. The number of vendors on a short list depends on how many adequate proposals there are. For example, if there were only six, you're fortunate; your challenge is selecting the most promising from those six. Beyond that, how many you select is a matter of how much time and how many resources you have available.

Sometimes, there are no adequate proposals submitted. This happens when your RFP was unfocused, ambiguous, or technologically too aggressive. Your organization's challenge is then to decide if you should revise and reissue the RFP or proceed based on the responses you received. If you decide to proceed, you select the proposals that at least address your primary concerns.

Site Visits

A site visit is more than a field trip. It is a chance to collect information on the technical aspects of the proposal response on the company, its management, and its future prospects. In other words, you are furthering the reviews you have already started. Expect to do a great deal of paperwork. Your group, for example, will probably be asked to sign nondisclosure statements if proprietary product or financial information is discussed.

Consider also keeping a *vendor tour log* of all site visits. After all, few people have perfect memories. With several long site visits occurring in a row, an accurate log can prevent the loss of important information. As fatigue and information overload set in, one site visit begins to blend one into another. A consultant-kept log serves as a refresher for those on the tour, improves accuracy in vendor evaluation and settles any disputes among team members.

Why visit a site in the first place? Site visits provide both objective and subjective criteria for evaluation. Visiting organizations provides a wealth of detailed information, as well as subjective impressions. However, the effectiveness of site business is increased by successful preparation. Take a list of questions with you to each site visit. This list should be based on the following:

- The details of your RFP
- Each of the qualifying vendors' responses
- Any general areas you want to cover

Answers to these questions help you revise your preliminary rankings of the finalists. To get answers, make sure you have access to senior management in larger companies and the CEO in smaller companies. You should also arrange time with the key managers who will be responsible for parts of the imaging system you need. These are the people you will need for troubleshooting; they should be able to address formidable problems.

Keep your investigation broad. While the primary goal of the site visit tour is to further screen qualified vendors for an existing RFP, you should also develop a list of qualified vendors for future projects. Now is the time to get as much information and as detailed a perspective as you can on all vendors.

Your team might also want to keep notebooks filled with impressions and feelings about each organization. While the primary purpose of the tour is to collect empirical data, you can never eliminate the human element. In most cases, imaging implementation relies heavily upon the vendor's personnel. Your instinctive feel about these people should be give consideration. In some cases, the proposals of competing vendors might be so closely matched in all objective areas that the real deciding factor is these feelings.

All subjective and objective information should go into a report summarizing all the visits and including vendors' responses to the questions developed before the trip. If you use a consultant along the way to log your vendor visits, record responses, or document impressions, he or she should also work on the report. You might want to have the whole team prepare the report, or you can present it to them as a briefing tool.

The Final Stages

Once all the reports from the site visits has been completed, it is time to have a final meeting of the full RFP team. This will give everyone an opportunity to query the visiting team to solicit unreported information and any personal impressions. By the end of the meeting, everyone should be satisfied that all useful facts and findings have been communicated.

All pertinent findings should be added to the decision matrix used to create the short list. Allowances should be made for such issues as

- Confidence in the vendor's management
- Key personnel
- Vendor's ongoing technical development strategies

Make certain that all factors, including these subjective ones, are accounted for in your formal report to management. The decision-making process behind your report should—and in some cases must—be auditable.

The final step is the meeting with senior management. As tempting as it is to just plunge ahead with this meeting, especially after all the work that has been done, you still must prepare. On its simplest level, that means gathering all materials needed to document your case as well as the request for a final pre-award meeting. You should also be sure to involve your legal department and your financial officers in this preparation. Both departments have the power to halt or delay the award if they detect irregularities, or if their own departmental procedures have not been properly executed. Working closely with all decision makers to see that they are well-informed and comfortable with the project again at this stage can prevent problems.

If your organization doesn't have much experience with imaging purchases, you might find it useful to review other imaging system contracts. Locating such contracts is easier than you might think; your local AIIM chapter probably can put you in contact with other organizations that have already purchased systems similar to yours. Existing contract clauses provide guidance on such issues as performance validation and penalties for defaulting on performance. Attorneys and consultants are useful during this process.

In the end, the purchase of an imaging system, from an organization's legal point of view, is just another contract. As soon as a purchasing decision has been made, be sure to clarify your course of action with your legal and financial departments, and after their approval, notify the vendors of your selection. The legal and financial procedures for awarding contracts vary greatly from organization to organization and from industry to industry. Your organization will almost certainly have formalized procedures for contract awards. Consult them carefully to ensure that the appropriate personnel are following those procedures.

Chapter

13

Training

Just as electronic imaging does not exist in a vacuum and needs to be looked at in the context of the computer industry as a whole, electronic imaging training needs to be viewed in the context of the broader training industry. Training is currently a boom industry. More and more, businesses are coming to understand the necessity of education. In 1988, the William T. Grant Foundation Commission on Work, Family, And Citizenship summed this proposition up by saying that most employers agree that this nation's future prosperity depends on the energy, flexibility, and creativity of a well-trained workforce that is knowledgeable, innovative, efficient, and dedicated to quality.

The Boom Industry

For must people, training has become a fact of life on the job—and one that is likely to take up more and more work time. In 1988, according to *Training* magazine, businesses with 100 or more employees spent $36.9 billion on formal employee training. In 1991, this figure climbed to $44.4 billion.

Those figures, moreover, are just the tip of the iceberg. We live in a training society where colleges and universities spend $80 billion per year educating students. These educational institutions spend more for qualifying and upgrading American workers than all other post-secondary educational institutions combined. The Department of Labor estimates that Americans spend a total of $210 billion on training each year.

Not surprisingly, almost one-third of these training dollars are spent on technical training. Increasingly, the American economy is being driven by technological innovations. The workplace itself is changing as a result. As new technologies become a substitute for human labor, tasks disappear or are recombined into fewer jobs requiring more skills. (Just try getting a job as a Linotype operator today!)

Economists talk about labor-intensive and less productive jobs being restructured into fewer capital- and knowledge-intensive ones. That means most people need to spend more time learning new ways of doing things to keep out of the unemployment line. The upside of all of this is that the line between management and workers is starting to blur. Decisions that used to belong to middle management now can be handled by lower-level workers. This trend toward technical training will continue. The Bureau of Labor Statistics sees the number of technical jobs increasing 21 to 32 percent over the next ten years.

Computers offer a bonanza for trainers and training organizations. At first, many human resource departments and trainers were computerphobic. As late as 1985, author Ronald Ribler was warning companies not to treat personal computers as a passing fad, writing, "Professional trainers no longer have the luxury of deciding they want nothing to do with computers. There are more of them than there are of us—and they reproduce faster!"

The hard part these days is trying to sort through the glut of computer-education opportunities to find the ones that make sense, especially for those for whom computer work is a tool and the central focus of their careers. Everywhere you turn these days—the sports page, the business page, even radio and television—you see ads for computer education programs. Sift through that pile of flyers and "junk mail" on your desk and you'll probably find the lion's share is from computer organizations offering hardware seminars, software seminars, one-day courses, correspondence courses, manufacturer-sponsored courses, courses taught by computer stores, as well as various self-teaching programs. You almost have to take a course just to understand who the players are!

Competition for training was not always so intense. Until 1969, manufacturers were able to make their systems available as a package that included all of their services: training, maintenance, and software support. However, as a result of an antitrust suit against IBM, manufacturers had to unbundle their services, making them available on an itemized basis. This unbundling enabled smaller manufacturers, vendors, and special software houses to compete against "Big Blue" and the other large manufacturers for each of specific services provided. Sweeping changes followed in the entire electronic data processing field as a result of the ruling, enabling smaller organizations specializing in one or two services to compete with the giants.

Computer vendors today often include training as part of their sales presentation. JVC Technologies, for example, will not sell a network to anyone without including training in the package. A spokesperson explains, "It is important to us that each of our customers remains a customer for the long haul, not just the original hardware sales. Our ability to offer a range of after-sale services—including technical support, system maintenance, and education—is crucial."

Many dealers say that their customers are demanding training. "Customers like to deal with a single source rather than purchasing a system one place and going to another for education," explains Judy Housman, president of Microsolve. "Even if resellers end up subcontracting some of their training offerings to a training specialist, it is to their advantage to deliver services on the premises."

Productivity can be increased through training. This is particularly true in the early stage of gaining familiarity with a system. At this point, the degree of improvement can easily justify training costs. The challenge for the organization is to create a test—some sort of indicator—that will keep your organization from becoming overtrained.

For employees, the harsh reality is that avoiding training is a good way to commit career suicide. Those accustomed to the world of typewriters might not like the new world of electronic imaging technology. However, sticking to the old ways of doing business guarantees personal obsolescence when a favorite piece of technology becomes obsolete.

"Users who are constantly enlarging their systems and installing applications software of continually increasing sophistication are a ready market for training services, yet they are becoming increasingly discriminating," Joe Shore, *Computer Reseller News* (February 12, 1992) wrote. Shore points out that not all training facilities are considered equal, quoting a microcomputer manager for a Fortune 1,000 retail company as saying, "We've found that almost every reseller offers education, but we've also seen unqualified instructors, inferior courseware, and little transfer of knowledge."

Shore's criteria for selecting a training organization are as follows:

- A professional teaching facility
- High-quality materials
- Certified instructors

Certification is the newest wrinkle in the training game. Earning certification often is a Herculean task. The theory seems to be that the harder something is to obtain, the more valuable it is. In other words, restricting the supply of certified trainers increases the marketability of the ones who are working. WordPerfect Corporation, for example, requires a videotape of an instructor teaching a class and the passing of a written test that can take 60 hours or longer to complete. So far, certification is not a prominent issue in electronic imaging training, however.

One of the reasons that many companies turn toward veteran talent for training is that they deliver what they promise. Too many novices make the mistake of overpromising in the first year. "The tundra is barren so they're tempted to plant something—anything—wherever they turn," says Beverly Geber in *Training* (September 1988). "They run the danger of failing to keep their promises, thus fatally damaging their credibility."

The State of the Art

The consensus among imaging vendors is that end-user education has been and continues to be essential for developing the imaging market. The basic problem, however, is economics. As the article "What Role Do Seminars and Conferences Play in Meeting End User Needs?" (*Imaging Technology Report*, October 1991) put it, "Conference and seminar providers have tried to fill that need [for education] and make money at the same time. In today's economy, it's questionable how many are profitable, and a number of seminars have been canceled or postponed." Some of the more notable cancellations have included the cancellation of Info 91 and the elimination of seminar schedules for Cohasset Associates and Rothchild Consultants (which went out of business).

The good news, however, is that many educational opportunities continue to exist. Not-for-profit corporations provide much of this education. They do not seek to lose money, of course, but have the capacity to absorb losses. Profit-driven firms also are continuing to offer courses. Indeed, seminars at conferences today are the best source of information on electronic imaging (see Figure 13.1).

Why are some seminars and conferences thriving while others are floundering? Those seminars catering to upper management might have a better chance of success. Tom Koulopolis, president of Delphi Consulting Group, told *Imaging Technology Report* that he "hasn't seen a dramatic impact on upper management attendees coming from large organizations." On the other hand, industry guru Robert Kalthoff explains that lower-level managers might be caught in a bind because "management is no longer looking at seminars as a way to educate. They're looking for a tangible payoff."

The audience for EIM seminars and conferences is evolving along with the industry. Many experts see an increased emphasis on experienced-based education over theory. Numerous users have graduated from basic concepts and now need more advanced training, such as help with systems management, applications, and figuring out how to apply new technologies.

This is not to say that introductory courses are passé. There is still a need for fundamental applications and solutions. Many first-time users cannot afford to miss a basic course.

New and more carefully targeted seminars are attempting to meet audience needs. The Association for Information and Image Management recently added many new conferences and sessions. Other educational institutions are adding imaging courses for the first time. For example, BIS Strategic Decisions recently has added an image compression seminar. (For more information on specific seminars and courses, see appendix A.)

CONFERENCE

In Just Four Days . . .
You'll Learn the Most Important Lessons
in Document Image Management,
from Acknowledged Leaders in the Field

MONDAY, April 5

9 a.m. - Noon,

1 Emerging Technologies

Join the members of AIIM's Emerging Technologies Advisory Group as they assess the impact of emerging technologies on specific areas of information systems. Examine the trend toward a multi-vendor environment. Discover the new directions that emerging technologies are taking. This session will examine technology trends in hardware, software, networks, input/output, storage, and information management engines. (G)

George Troell, Session Moderator
Ford Motor Company

Speakers to be announced

2 Introduction to Electronic Image Management

Discover the technology and concepts involved in electronic image management. Learn the difference between data and documents. Understand the media, hardware, and software used in electronic imaging. Discuss the varying points of view on key issues such as legality, standards, systems integration, and workflow. Find out how to get started. (D, E, G, H)

Don M. Avedon
Avedon Associates, Inc.

3 Introduction to Micrographics

Become acquainted with micrographics terminology, characteristics, production, user systems equipment, and standards. Active and extended data storage issues, coupled with quality assurance requirements will receive specific attention. This session will include a question and answer period. This is an interesting session for producers, users, managers, or buyers of micrographic products and systems. (C, D, E, G)

Whitney S. Minkler
MSTC

Jerry Vilhauer
MSTC

9 a.m. - 10:15 a.m.

4 Beyond Workflow: The Large Files Problem

Traditional workflow approaches to image system design concentrate on simplifying the flow of information. While important, this approach ignores the significance of engineering the image environment to optimize the powers of the professional person working in a large and complex file. Examine some of the visual cues and clues that can be replicated or replaced in image implementations for large-file applications. (G, H)

Ray Grenkie
Coopers & Lybrand

5 The Economics of Standards for Imaging

Standards development has become a big business in the United States. The acronyms, scope of work, and structure for the development of standards in the U.S. and abroad is complex. Learn who is doing what in standards, and why.

KEY TO DISCIPLINES A: Applications and End User Experiences B: Large Format Drawing Applications C: Micrographics D: Project Management and Methologies E: Standards, Records Management, and Legal Issues F: Strategic Focus G: Technology Features and Functions H: Workflow and Business Process Reengineering

Figure 13.1 A few course descriptions from a recent AIIM conference.

Suggestions Before You Shop Around

Every organization needs to keep in mind four basic questions before hitting the training trail:

- Who will you train?
- Who will do the training?
- What do you expect the training to accomplish?
- Where will you do the training?

Who will you train?

The question of who to train is harder to answer than you might expect. Oldtimers, accustomed to tried-and-true ways of doing things, might not want to be trained. Also, not everyone shares the same aptitudes. This has nothing to do with education level or status within the company; the fact is that some people are more willing to pick up computer skills and better at it than others.

Many training experts say that it is okay to mix new and existing employees, and those of different aptitudes, in the same class. In fact, many educational models stress the idea of peer training—having more skilled individuals tutor their less skilled peers. The ideal peer trainer, the experts say, is someone who

- Is motivated by his or her work as well as pay
- Likes or can endure the attention of being a peer tutor
- Is knowledgeable, articulate, and self-confident

Who will do the training?

You basically have two options regarding who will handle training: in-house staff or outside consultants. The problem with relying on in-house staff, however, is that EIM technologies often are so new that your staff might not be abreast of all of them. Added pressure for quick mastery of electronic imaging systems also comes from cost-conscious managers eager to recapture high capital costs. After all, electronic imaging systems only start saving money when people can use them.

Before hiring consultant trainers, however, organizations need to understand consultants and their roles. Consultants offer a variety of services, ranging from high-level strategic planning and organizational diagnosis to writing computer system documentation and training manuals. Many prefer the more glamorous and lucrative assignments such as advising high-level decision makers (you, your boss, your boss's boss) or facilitating task forces and committees. Consulting groups can provide you with a single worker, a pair of workers, or a whole team of workers, depending on your needs. There is also no shortage of credible one-person shops.

You might be confused by differences between an independent contractor and a consultant. The differences are semantic. Contractors generally work for themselves, independent of any umbrella consulting firm. Contractors are often technical experts, such as software designers, evaluation specialists, scientists, or artists with specific skills. These contractors, who usually have low overhead, are generally paid per hour, and paid directly. Consultants, by contrast, are often salary-earners, members of a consulting firm or group who are paid an hourly or per diem rate in an attempt to cover overhead and make a profit.

Finding a good consultant is an art form similar to selecting a physician who is right for you. Before hiring an outside trainer, you should examine and evaluate the following:

- *Reputation.* Reputable consultants should tell you exactly what they did on each assignment and when they did it. They should also give you references to contact. Some training firms are stronger in some areas than in others. Also, consider a firm's overall breadth, depth, longevity, and amount of repeat or referral business (a good range would be 80 percent or more).

- *Business practices.* Find out about personnel philosophies. Some consulting firms offer only part-time consultants who are employed only for the duration of a specific project. Other firms hire full-time employees experienced in the specialty of the firm. Such firms can generally offer you replacement consultants of similar quality in case a consultant assigned to your project leaves the firm or becomes ill. Support personnel is another important issue. A good firm will have adequate support staff. This is especially true with firms involved in electronic imaging, where the lines between professionals and support staff are blurred.
- *Facilities.* Pick a training firm with adequate software, printers, and computer equipment. This is especially true in electronic imaging, which is based upon the mastery of this equipment. You should get a sense that your trainers use much of the equipment and procedures they talk about on a daily basis.
- *Personality and personal characteristics.* Of course, you will want trainers you can trust. Beyond that, however, look for someone capable of providing "good chemistry." In other words, you want a trainer who is pretty much like your current staff in personality, who will fit into your corporate culture easily, and who can relate to the people you want to train.

Many companies are unaware of just how extensive consultative services can be in the educational arena. You can turn to a consultant to contribute technical or specialized input to course designers on your staff or to develop an entire training marketing plan. Some firms use consultants to evaluate or assess an existing program. Other consultants can create a course from a review of current documents or new technical developments. For example, a consultant can review current successful applications of a new technological application to create a new generic course. Consultants also design entire multicourse curricula, prepackage workshops, write course manuals, and teach courses themselves.

Human resources also turn to consultants for some of the more esoteric aspects of training, such as designing a test or evaluation system. They can provide research, prepare instructional videos, slides and transparencies, or design and produce instructional charts and job aids. Some of the more business-oriented training firms even broker services such as printing, packaging, and mailing.

The trick, of course, is finding the right consulting firm for your organization's needs. Some groups offer experienced advisers who dole out advice for a high per-hour fee. Others supply workers who function as helping hands, working at low rates to build up expertise.

Considering the amount of work involved in selecting a training consultant, the idea of self-training in EIM seems tempting. Such a shortcut might seem like a welcome relief to anyone who realizes just how much time and money can be involved in using outside trainers. However, the real costs of this type of program are higher than you might expect. The time commitment for self-teaching is enormous. You will have to spend much more time to reach the same results that a good trainer could provide. This is especially true for people who are not computer-literate and have to get used to some of the basic concepts of using computers before they can master the more advanced aspects of electronic imaging.

On the other hand, preparing yourself for learning about electronic imaging does make sense. A good way to start is to visit the Association for Information and Image Management resource center. That way, you can really dig into their literature. If a trip of that kind doesn't make sense, you could start by studying the books and other publications listed in appendix A.

Communicating with vendors as well as consultants is one of the best ways to educate yourself. Vendors tend to have a high respect for knowledgeable individuals. Good vendors seem to prefer talking about the latest developments in the field than in making quick sales. Make at least four or five site visits to vendors before making a purchase. That way, not only will you get a sense of the options available, you'll also end up with a crash course on state-of-the art methods.

Outside trainers ultimately fall into two categories: suppliers and third-party. Third-party trainers often can be found locally. Unless you're lucky enough to be in the same city as a training center for a major supplier, third-party trainers offer a cost advantage by eliminating most transportation costs. Third-party vendors also often tend to offer lower prices than vendors in order to attract business.

There are, however, some disadvantages to using a third-party training supplier. They're rarely as knowledgeable about a single system as a supplier is. This is especially true of cutting-edge technologies, where third-party trainers must become students first. Their equipment configuration at an off-site location might not be a good match to yours.

Regardless of the type of training organization selected, look for one-on-one attention. Small teacher/student ratios tend to be a good sign that this type of training can take place. Individual instruction is important to many students, who learn quicker and retain more when they are given this kind of attention.

What do you expect to accomplish?

Above all, training is a communication tool. That means that one of the hidden benefits of training can be improved communication. Because of the nature of training interaction, training becomes a way of forging a bond between trainees. Good training supports individuals who request clarification, ask probing questions, and volunteer information. In other words, training is a tool that can create an improved climate for all aspects of human relations. Individuals with fine-tuned communication skills end up working more effectively in task forces, committees, and other work groups. Meanwhile, training is a way to build

- Skills
- Attitudes
- Knowledge

The most basic skills in the area of electronic imaging involve mastering the basic equipment, both hardware and software. For example, a typical course might teach technicians the art of scanning or how to add images to databases. More advanced courses start to cross over from specific skills into general knowledge. Besides increasing the general knowledge about the technologies involved, training can also help pass on information about a company and its customers. Employees frequently need to understand where they fit into their organizations. Training can reiterate their standing and what is expected of them. It can establish standards of professionalism that carry into the workplace.

Employees learn a great deal about how their corporate culture really works by the attitudes found in training sessions. Is leadership encouraged? Are they treated with respect and compassion? What goals are supported? Certainly, these attitudes are evident in other areas besides training. Training, however, is a hypersensitive arena because of the assumed connection between training and promotion.

Where to train?

Of course, the least expensive way to train is to use your own facilities as a training site. You can cut travel expenses, and give trainees the advantage of actually working on the equipment they will be using. This method, however, can be extremely disruptive. Trying to train people on the same equipment that people are using can be tragicomic. As every high school science student knows, "No two bodies can occupy the same space at the same time." Sometimes, however, organizations end up learning this lesson the hard way. At the very least, using your own site as a training ground requires special coordination.

Many organizations, moreover, have a middle ground between outside consultants and regular departmental teachers. Miniaturization of computer capabilities and the spread of user-friendly technologies has caused data-processing operations to become more dispersed. These days, data processing operations serve multiple roles as line departments, research and development centers, and service and training providers. Some employers have created a new type of department, the information center, to manage this diversity. The information center serves as an intermediary between the database managers and end users. Sometimes they provide classroom training, but more importantly they are a hands-on laboratory cutting across departmental lines to assist people with hardware integration and software application problems.

Other companies, meanwhile, are working to decentralize their training facilities. After all, many of the advantages of in-house training, including reduced travel time and expense, are eliminated if you have to ship off to a corporate training site out-of-state. Also, high-tech facilities tend to have unique and immediate training needs. For example, why should a biotech facility have to share training facilities with an electronics division?

In reality, many companies end up mixing centralization and decentralization. Alcoa, for example, emphasizes individualized training at its Alcoa Laboratories work sites for managers, technical professionals, and laboratory technicians. Classrooms at Alcoa Laboratories are made available to engineers from other Alcoa locations when space is available. Alcoa Laboratories also operates two resource centers offering media-based courses for independent study. These centers, too, are open to other Alcoa divisions.

Other companies have centralized planning, but execute training in a decentralized way. For example, Federal Express has three training divisions. Its Air Operations division oversees flight operations training at a centralized Memphis facility, but often delivers courses on-site. Domestic Ground Operations is responsible for training all operations not covered by Air Operations, such as couriers and customer service representatives. This division also develops programs in Memphis, but executes many of them on-site. Human Resource Development covers career development.

Training often involves cooperative ventures with educational institutions. Colleges usually can provide qualifying courses (broad-based background courses) as well as adult-education seminars. The colleges benefit from having needed students; the businesses benefit by not having to add extra staff.

Some companies have gone as far as to forge direct links with colleges and vocational schools. Cincinnati Milacron provides training and technical assistance on robotics systems to the faculty of Piedmont Technical College in Piedmont, South Carolina. As a result, the local faculty are able to keep abreast of Cincinnati Milacron's training needs. Digital Equipment Corporation (DEC) has taken this concept one step further, supporting a nationwide series of programs to train computer service technicians at local schools. DEC trains the faculty, provides equipment, and allows schools to customize the curricula.

Colleges, in turn, can offer customized training. Piedmont Technical College, for example, created a special one-year maintenance program for

Monsanto. Monsanto provided job analysis information, which Piedmont used to develop a curriculum. Monsanto reviewed the curriculum and promised to hire all the program's graduates.

Unfortunately, at the time of the writing of this book, very few colleges offer courses in electronic imaging. Greater awareness of the rise of electronic imaging, however, should translate into more courses. Meanwhile, data processing courses would do well to incorporate electronic imaging information.

Advice for Trainers

Most American adults seem to favor the visual sense as their primary source of information collection. Do not neglect the importance of stimulating the other senses, however. The rule of thumb in training has always been that retention is directly related to how many senses are involved in learning. To really learn something you need to engage many of the senses at once.

That's why lectures tend to be boring and ineffective. Most lecturers just use sound, droning on and on, often in a monotone. Reading dry text is another ineffective way to try and learn anything; in fact, it might be worse than listening to a boring lecture. Either way, you're liable to end up asleep. On the other hand, supplementing these materials with great visuals, demonstrations, discussions, and other sense-involving activities will turbocharge learning. Hands-on practice is one of the most effective ways to learn things. All of the senses tend to be stimulated.

Never underestimate the power of graphics as a training tool. As the old saying goes, "A picture is worth a thousand words!" To be more precise, graphics offer the following advantages:

- Graphics force the teacher to focus. Even the most conceptual teaching point can be turned into a concrete example by use of a good graphic.
- Graphics decrease the possibilities of misunderstanding. It's a lot easier to follow a map than try to remember someone's instructions.
- Graphics open the door to more sophisticated instructional media. Graphics first used for lectures or desktop publishing presentations can be recycled for use in computer-aided instruction, computer-based training, and video.

Thanks to computers, working with graphics today is easier than ever before. Many computer programs today feature draw programs. Those with artistic skills and talents can use these programs to create original art. Virtually anyone, moreover, can take advantage of clip art. You can acquire clip art as part of many software programs, order specialized catalogs (such as for baseball or medicine) separately, or pick up public-domain or shareware versions through electronic bulletin boards.

Ultimately, however, graphics are only one effective teaching tool. Training at its best is a creative art. While there are no hard and fast rules for what makes a good trainer, trainers have, over the years, developed techniques that work especially well for adult education.

An effective training presentation, for example, will review the course requirements. In other words, the trainer must make sure that the trainee understands what the new task or situation requires. This can begin with a presentation of the "big picture." Most employees like to make a contribution to the total work. Describe the totality of that work so that trainees can see their roles in it.

Presenting background information is also useful. Although American businesses are often historically myopic, many employees appreciate acquiring the prerequisite information he or she lacks. If you cannot provide all of that information yourself, be sure to present material that points trainees in the direction of people or documents that will be able to provide that information.

In almost all cases, preparing a lesson plan is a necessary step. Training is not a field where "winging it" usually works. A lesson plan states what the trainer needs to do to help the trainee learn. A good lesson plan will be designed from the trainee's point of view, offering specific objectives to be reached. The best lesson plans break down objectives into very small steps, which are easier for trainees to master and invite more feedback. The trainer should also plan *feedback exercises*, the updated version of quizzes. Feedback exercises offer trainers a way of judging whether trainees got the message.

The lesson plan also works well as a managerial preview. Some prospective clients insist on a lesson plan as part of the selection process. Once a vendor is selected, the lesson plan can work as both a preview for prospective students and the basis for evaluation. Many trainers hand out their lesson plans to provide trainees with a "roadmap." Trainers also often hand out other paper items, including the following, to help demystify the training process:

- A flowchart of a process
- A job aid or card containing key operating instructions
- A form for trainees to list personal goals for this training

Trainees should be encouraged to keep these handouts as a first step in jogging memory after the class. The idea, of course, is that training is supposed to stay with students. Follow-up obviously should be incorporated into the lesson plans. While no one wants to create test anxiety, the reality is that training classes, although often informal, are important. Trainees need to know up-front about any follow-up they will be expected to perform. At a minimum, the trainer needs to make it clear when this will occur and what kind of monitoring or checking will follow. Better yet, trainees themselves might be willing to guide trainers toward the best kind of follow-up and its timing.

Trainers should do everything possible to make the training experience meaningful in its own right. That means approximating work experiences as much as possible. Be sure that the terminology used in training is the same as that used on the job, and use the same manuals as are used on the job. As much as possible, equipment should duplicate what trainees will use in their jobs. Part of a person's adaptation to new work is adaptation to dimensions, such as the height of the installation and the elbow room around the new machine or machines.

As a rule of thumb, training sessions should not last more than two or three hours at a time. After that time, trainees become fatigued. Facilities should also be selected that add to trainees' comfort. Rooms that are comfortable, clean, and free of interruption are suitable sites. Trainers can add to the atmosphere of comfort and professionalism by ensuring that all materials are clean and organized.

Friendliness goes a long way in the training business. A friendly atmosphere can be enhanced by simple amenities such as serving coffee and introducing all trainees to each other. The best trainers make their students feel special about what they are doing. A nice atmosphere is more than common courtesy; it's good business.

Computer-Aided Instruction

Just deciphering the alphabet soup of jargon can be a barrier to using computers as educational tools. For example, computer-based training, also known as *CBT*, could be defined as an interactive learning experience between a learner and a computer in which the computer provides the majority of the stimulus, the learner must respond, and the computer analyzes those responses and provides feedback to the learner. In other words, a person uses lessons presented on a computer, which also corrects errors. CBT programs often use the telephone or mail to interact with live teachers who also work with the computerized information. IBM's Multimedia Division has calculated the relative retention rates of various forms of training. They found that reading has a 10 percent retention rate, classroom presentations 20 percent, classroom presentations with visual aids 50 to 60 percent, while interactive CBT produced a 80 to 90 percent retention rate. Obviously, CBT has a bright future.

Computer-aided instruction (*CAI*) is another term used for the same process as CBT, but with slightly different technologies. Usually, CAI programs are more self-contained. You don't have to interact by mail or telephone for a lesson. Lessons come preprogrammed, like the videotape you put in your VCR at home. Technologically, CAI can be broken down into compact disk-interactive (*CD-I*) and digital video interactive (*DVI*) technologies, which are related to the storage medium used in the process.

Both CBT and CAI offer the following advantages:

- Immediate feedback to trainee responses
- Simulation of situations where skills, knowledge, or behaviors that have been learned through other training methods can be exercised, practiced, or tested
- The act of using computers to present data training itself is a form of computer training
- Students cannot proceed to the next learning step until they have mastered the current step

Despite these advantages, however, no one would suggest that computers take the place of books, films, and live instruction. There is no such thing as a perfect solution for any training problem. However, CBT might be the best solution under the following circumstances:

- Trainees are comfortable using computers
- Good training programs are available
- Your computer system can handle this type of program
- Your organization has some way of monitoring the success of this type of program

Marc Di Stasi, who produces training videos on a variety of topics, including computer-related topics, says "Video is taking a more and more active role in training. It's not a medium just for entertainment and news any more. However, those two elements are utilized in some of the more successful videos, ones that are both instructional and interesting." Di Stasi offers the following tips for anyone planning to create a training video (or any other form of training):

- Carefully design your message (e.g., "How to Operate *X*").
- Describe your audience (e.g., "*X* Operators, Managers, and Maintenance Personnel").
- Determine objectives. Typical objectives would be to improve productivity, reduce down time, or provide a safer work environment.

"These three elements are then worked into your script," Di Stasi explains. "Your script should start off by telling your audience what you are going to instruct them on, then the program should go through the process of the actual instruction. The conclusion should reiterate what was presented to your audience. Along the way, every element of the program should be coherent, including the scripted words, graphics, and sound."

Not every company, of course, has in-house production facilities or staff. Many companies turn to outside production houses, which have their own staffs of writers and producers with educational as well as video backgrounds.

"A video production house can assist you in writing a script that will help your audience clearly understand the information presented to them," Di Stasi concludes. "Also, they can help avoid any pitfalls when it comes to production. They understand the production process and how it relates to the expression of your instructional information. For example, in some situations special effects are helpful. In other cases they are distracting."

Videoconferencing and Teletraining

Teletraining uses telephone lines to carry a training program between an instructor and a group of trainees. This instructor teaches from an office, classroom, or conference room, using an audio-talk system (audio link) to ensure two-way communication. This system tends to work better for knowledge-based training than skill-based training. The use of fax machines, however, can increase the ability to send complex bits of information from the students to the teacher and vice versa. Generally, teleconferencing is recommended for short (one- to three-day) courses taken in one-hour increments to students spread out over a wide geographical area who need access to information quickly, and for organizations who have site coordinators and technicians who know how to teleconference. The biggest drawback of teletraining is a lack of eye contact with instructors.

Videoconferencing adds the dimension of satellite transmissions. This technique requires the teacher to learn television skills, including the following:

- How to speak and look directly into the camera
- How to adapt documents and graphics for broadcast
- An understanding of production issues

In general, videoconferencing makes sense when urgent need to educate large numbers of employees can justify the costs incurred because what they need to know is critical to the business. Educating large numbers of people this way reduces the per-student costs. In fact, many videoconferencing advocates argue that videoconferencing is the training solution addressing situations where per-employee costs for training-related travel have gotten out of control. Videoconferencing works best when there is a need to respond quickly to changing information.

Videoconferencing centers are appearing all over the world thanks to increased customer demand. From 1990 to 1991, the International Teleconferencing Association (ITCA) found that the industry grew from $365 million to more than $510 million. The ITCA predicts continued growth."

"As technology improves and costs for conferencing equipment decreases," Stephen Hiscott, ITCA manager, explains, "more and more businesses are discovering videoconferencing is a cost-effective alternative to travel."

A one-hour video conference between New York City and Los Angeles should cost approximately $800, somewhere between the cost of a coach and a first-class roundtrip airline ticket. Savings are even more pronounced on the international level. It would cost $5,000 or more to fly from Los Angeles to Tokyo and stay at a hotel. A one-hour international videoconference averages $3,000.

Some businesses have set up their own private videoconferencing centers. It costs more than $20,000 to equip such a center with video screens, electronic pads for transmitting data, cameras, and audio systems. Smaller businesses lease rooms on a hourly basis at one of the more than 1,000 such videoconference centers available nationwide.

So far, teleconferencing has not been used extensively in electronic imaging training. However, the potential for its use is strong.

Training as a Tool for Growth

At the very least, human resources management requires policies that avoid wasting time, money, personnel, and work on tasks that don't add value. No company has unlimited funds. Training programs must compete with other operations for a fair share of corporate resources. In other words, corporate training needs to justify personal growth by its impact on the productivity, profit, growth, and development.

Managers can ensure maximum benefits from training if they follow these basic guidelines to make training quick and effective information:

1. Provide equal opportunity. Training is often viewed as an employment opportunity, as a precursor to promotion or a raise. In today's litigious environment, it can be a factor in litigation by an individual against a corporation. You can avoid serious problems and get better results if you guarantee equal opportunity in training, equal opportunity to get into training, and equal opportunity to succeed during and as a result of training.
2. Be prepared for skepticism. The best way of convincing doubters at all levels is by proving that good training pays off. You can win friends among senior management if you measure training's contribution to corporate growth by its effect on quality and productivity. You should also tie training success to your company's individual formulas for productivity and quality.

3. Design quality programs. It's always easier to prevent mistakes than to undo them. To ensure your programs are the best possible, make a concerted effort to seek input from several different kinds of people who have a stake in the results of every stage of the training you propose.
4. Think small. The best way to design training programs is to create small sections with specific objectives for specific learners. Trying to be all things to all people will result in spending too much time in development and will diminish the impact of the delivery of training.
5. Think individual. Effective programs design and deliver training around job experience. One-size-fits-all programs rarely make sense. As much as possible, remember you are designing training for individual adults.
6. Be creative. Don't be afraid to vary your presentation methods. Consider using charts, posters, videotapes, structured discussion groups, and demonstrations.
7. Preview and customize. You might want to consider customizing a generic commercial product, such as a videotape, but don't buy it without seeing it. Avoid products that are padded with irrelevant material, and be sure that there are logical interruption points that allow you to insert your own information.

Chapter

14

System Evaluation and Maintenance

An electronic imaging system is an organic whole. After you have initiated a system, your attention turns toward evaluating it, maintaining it, and helping it grow.

Evaluation

After your imaging system has been installed and been running for a few months, you should undertake a formal evaluation. This evaluation process looks at both the project and the process. It provides needed feedback to everyone who participated in the project and helps guide them towards detailed refinements and additions. This evaluation process should cover three areas:

- Project evaluation
- Vendor evaluation
- User evaluation

Project evaluation

The project evaluation requires input from senior corporate management, system users, and managers as well as from members of the original project team. While users and managers will want to focus on evaluating the work of the project team, the project team should join everyone else in looking at hardware and software performance and an analysis of the vendors. The evaluation of hardware and software performance is also known as *system evaluation*.

Performance can be reviewed in terms of how well contractual guarantees met requirements. Questions to be investigated include the following:

- How quickly are images retrieved and displayed?
- Are all necessary images and documents online?
- How well do the backup and disaster recovery procedures work?
- Are all ties to existing MIS systems functioning properly?

This review process need not focus only on system shortcomings. You might find that your company has made revisions in work processes or system hardware placement. Such improvements do not necessarily reflect poorly on your project team's earlier work. As system users and managers gain experience, improvements are inevitable.

The basic underlying question for project evaluation is, how successful was your imaging system implementation? You review your project goals to find the answer. Often you will be pleasantly surprised to find that most goals have been surpassed.

Vendor evaluation

Vendor evaluation after the system is up and running focuses on the following:

- Hardware maintenance
- Software support
- Training

You should be looking at both technical questions and personnel issues. Technical questions include these:

- Do system elements perform properly?
- Is your system regularly maintained?
- Has acceptance testing uncovered any performance or functional shortcomings in software?
- Did the vendor offer the training your organization needed?

Personnel questions include these:

- Have you found the vendor's software support staff informed and responsive?
- Is your staff comfortable with the vendor's staff?
- Have the trainers been accessible to answer problems?

User evaluation

The goal in a user evaluation is to find out why some users are more productive than others. Finding the answer to this question will lead to improvements in productivity. Sometimes they also lead to improvements in the system.

Post-implementation interviews usually are the best way of handling user evaluations. These interviews can

- Identify emerging problems
- Demonstrate a continuing commitment to listen to worker needs
- Solicit system improvements
- Provide a basis for analyzing productive and nonproductive workers

Interview techniques can range from formal surveys to informal discussions. In the end, you should prepare a report on the information collected.

Estimating the Total Cost of Maintenance

Many of the real costs of maintenance are hard to detect and harder to track by the relatively inexperienced system user. The costs of maintenance can be divided into two classes:

- The cost of fixing your system when repairs are needed
- The cost of your system not being operational while it is being prepared

These costs are not always obvious. The readily visible expenses of maintenance calls are easier to document than the less visible costs of system inactivity and replacement of lost work that hadn't yet been saved when the malfunction occurred. Every time a computer breaks down, it costs your company money.

The good news is that computers don't break down the way they used to. In the old days, when one terminal broke down, a whole system would fail. Now, however, a single terminal can break down without affecting the rest of the system. The bad news is that vendors are less willing to provide free service of any kind. Indeed, as hardware service revenues continue to decline, vendors are looking to their support options to take up the slack. Just about everything is, or will soon be, chargeable. The additional consulting and all the other little extras (such as technical telephone support and applications help) that large users have gotten free in the past as part of standard maintenance contracts are largely chargeable items today.

One thing hasn't changed, however: downtime is an annoyance. Still, some areas see more downtime than others. For example, electronic publishing systems have undergone technological improvement at a rate that outstrips the desire for maintenance. By the time a persistent problem is recognized in a delivered system, suppliers are often too busy with next-generation replacements to respond.

How much downtime is acceptable? There is no standard answer. Some amount of downtime is inevitable. To those impatient to pay back their system, any downtime is too much. The fact that downtime lengthens payback periods should be factored into all financial planning for maintenance. Calculate the effects of downtime as a function of labor-hour cost and payback period extension.

For example, suppose you have a $525,000 system with savings enabling a three-year payback. In other words, the system as configured, with its processors, disks, and output devices, would realize a net savings of $175,000 per year. Suppose you have a maximum of 37,500 labor hours to do it. You must, therefore, save a minimum of $14 per labor hour.

If a terminal is down and a worker is idle, your company is losing the $14 per hour needed for payback. Worse, your company must still pay the operator his or her base salary, let's say $22 per hour. The real cost of downtime in this example, then, is about $36 per hour. This cost would climb considerably if your whole system went down.

Maintenance Strategies

There are three basic maintenance strategies:

- Prepaid full-service contracts with a supplier of preventive maintenance providing repairs as needed
- Time-and-materials agreements with maintenance suppliers
- As-necessary maintenance from suppliers of convenience

Full-service contracts

Prepaid contracts can take many forms. For example, many companies have a written contract with a supplier agreeing to preset prices. Many of these contracts run at least a year and are paid monthly or quarterly. Prices usually remain constant for that period. Sometimes, however, you might want to add new equipment to existing contracts. Costs then go up. Many such contracts also include clauses describing out-of-coverage items, such as user-caused damage. In those cases, clients are billed charges above the fixed prices. Other times, contracts spell out premium labor rates for after-hours repairs.

These additional clauses are important because from time to time most organizations need some maintenance activities that aren't covered by prepaid full-service contracts. These needs might arise from severe emergencies that existing staff cannot handle. In many cases, these emergencies take place outside of the usual first-shift, Monday-through-Friday periods. Suppliers usually charge a premium for sending extra support at this time.

Prepaid full-service contracts effectively are like paying an insurance premium in advance for the time period covered by a contractual agreement. Unlike most insurance policies, however, you negotiate a service contract. In this case, you negotiate the contract with the maintenance supplier. The price of this contract is fixed and known, but the cost of maintenance actions is who-knows-what!

As a rule of thumb, full-service (prepaid) contracts alone typically cost 9 to 15 percent of the acquisition price of the hardware items under contract each year. Software maintenance (troubleshooting as well as updates and modifications) is usually an additional expense. Thus, many companies seek cheaper solutions.

When you have a malfunction under a full-service contract, the supplier sends appropriate specialists out to your site to

- Troubleshoot the malfunction
- Obtain and install the replacement parts
- Verify that the system has been restored to productive operation

The supplier bears some costs, including the following:

- Transportation (both for the technician and the parts)
- Labor
- Parts and materials

This supplier, however, is not responsible for any of the costs of your downtime. The suppliers do not have to cover losses that occur while you're waiting for the technician's arrival or during the repair.

Full-service contracts, however, do not cover problems due to negligence, such as spilling coffee on a keyboard. The buyer then is responsible for decreasing the risk of such accidents. For example, a company can institute a no-eating, no-drinking policy for computer areas. Buyers can also decrease the need for service by routine preventive maintenance. A system should be inspected and potential problems spot checked occasionally. Suppliers also benefit from such corrective actions.

Time-and-materials agreements

Time-and-materials agreements cover only the supplier's minimum costs for bothering with the agreement. Under this system, each visit is priced separately. You pay for labor, parts, and supplies as needed. The advantage of this system is that you are guaranteed a supplier who has the parts your systems need without having to pay a large fee if you never use their services.

As-necessary contracts

As-necessary maintenance is the riskiest strategy for a user. Time-and-materials agreements have response times built in, but as-necessary maintenance is unpredictable. There are no timetables. Unlike any of the other types of services, there is no guaranteed access to supplies and spare parts.

This agreement is also known as "going bare." As tempting as that provocative name is, it is not a preferred approach, especially for the neophyte. You need to be able to cover your losses if your luck goes sour.

Maintenance contractors, on the other hand, are quite able to afford the ability to assume risk. Maintenance contracts are a very profitable business for suppliers. There are also some businesses that are maintenance suppliers exclusively. These are run by independent practitioners who depend upon their general maintenance expertise and their ability to acquire equipment documentation and spare parts as needed. Original equipment suppliers, however, control most of the maintenance-providing market.

Cost Model

The easiest way to estimate maintenance costs for a full-service contract is to use the following model:

Downtime Costs + Full-Service Contract Costs + Out-of-Coverage Costs = Total Costs

If a contract is new, you have to make estimates based on manufacturer recommendations or advice from other customers. Once you have a track record, you can base service costs on past performance.

As an example, let's use the $525,000 model with a three-year payback period and a 10 percent ROI. A full-service maintenance contract could easily cost 12 to 15 percent of the original purchase price per year. Accordingly, you might set aside $189,000 for planned and contractually covered maintenance activities (12 percent over three years), or $63,000 per year.

The full-time service coverage can include unexpected expenses. Additional maintenance costs will occur if you have existing equipment currently not within the paid care of your maintenance contractor. In other words, you can face a surcharge. Even the cost of a qualifying inspection will be borne by the user. New items added after the inception of full-service coverage might be surcharged to accommodate the administrative costs of contract modification as well as additional prorated coverage.

Estimating out-of-coverage costs is difficult. Still, you have to put aside money for emergencies and accidents. In the absence of personal experience, other customers with similar systems are the best sources of just how much to budget in this area. Certainly this budget will be much smaller than your full-service contract. Suppose you estimated 1/10 of your service contract costs. In the example discussed earlier, your figure would be less than $7,000 each year.

Downtime estimation is also tricky. You might have a good idea how much your downtime will cost (let's say $36 per hour as before), but you can't know in advance how many hours of it you will face. Here again you must estimate. Suppose a reasonable figure of 2 percent; you're looking at a downtime loss of $2,700 per year, for a total maintenance cost of $72,700.

At this point, the full-service contract and contingency activities cost outweigh the costs of downtime maintenance. However, while the costs of maintenance contracts increase only slightly as the system ages, the costs of downtime go up dramatically. If, however, the loss due to downtime starts to catch up with the costs of maintenance contracts, that is a signal that it's time to change systems. A beefed-up maintenance contract will not suffice.

Warranties

A warranty for a computer system is much the same as the warranty you would buy for your automobile or a major appliance. It is a protective step taken by the manufacturer or supplier, giving the buyer low-cost maintenance or free service for the period immediately following the purchase. Usually, a warranty covers all parts and labor for a reasonable timeframe and frequently gives partial coverage on certain types of problems afterwards.

The manufacturer is giving you time to verify the integrity of its products and to protect against basic flaws that would almost certainly show up within the first few weeks after installation. The warranty period, however, does not cover problems caused by routine wear, nor is it a substitute for regular maintenance.

There is no such thing as a standard warranty. Because of this a buyer should ask the vendor questions such as these:

- When does the warranty period start? (Some start at the delivery, others at the installation.)
- When does the warranty period stop?
- What restrictions apply?
- Do you have to register your equipment?
- Do you have to take action in any other way to activate the warranty?

In addition, the buyer should compare the warranty with any maintenance contracts to see if there are any gaps between the end of warranty and the beginning of regular maintenance coverage. Plans should be made for dealing with breakdowns during that period.

Software warranties are uncommon. In many cases, they are not available, even upon request. However, faulty software usually is easy to replace. You can also prevent problems caused by computer viruses by careful testing and by avoiding public-domain software.

Third-Party vs. Vendor Maintenance

Third-party maintenance suppliers have two advantages over vendors: they tend to have lower prices and offer convenience. In many cases, manufacturers are unable to supply maintenance to remote locations. In those cases, special deals are struck between third-party suppliers and manufacturers. The third-party-maintenance provider might pay the supplier for the right to be an "authorized" maintenance supplier for that vendor. In turn, the original supplier will provide services such as the following:

- Spare parts
- Training
- Documentation
- On-call expertise

By the time your system has been in operation for more than a year, many third-party maintenance services will probably have heard about it. The more aggressive ones will seek bidding rights on your annual contract. This is good news because competition reduces prices. The best way to evaluate these services is to look at areas such as the following:

- The competence of the maintenance technicians
- Staff size
- Access to troubleshooting documentation
- Access to spare parts
- Test equipment
- Affordability
- Reliability
- Business stability

Competent technicians with access to documentation are key to reducing downtime due to maintenance. It will not help your organization to rely on a contract that saves even a substantial portion of annual maintenance costs if you wind up doing business with less-qualified technicians who take much more time to fix your system, or who take much too much time waiting for spare parts.

In the final analysis, however, third-party maintenance contractors might be a dying breed. According to *Information Week*, by 1991 third-party firms had only 11 percent of the market for maintaining mainframe computers, as opposed to mainframe manufacturers' 84 percent of the market. Third-party maintainers are facing hard times, analysts say, because mainframes do not fail as often as they did in the past, a fact that renders price maintenance contracts almost superfluous. They are also losing ground to competition from giants. In 1986, IBM triggered a price war with their joint service amendment that essentially offered buyers a 20 percent discount on maintenance. Meanwhile, specialists, such as TRW Customer Service Division and Intelogic Trace have moved into microcomputer maintenance.

Original suppliers tend to be the most knowledgeable and expensive sources of maintenance. After all, they sell more of their own machines than anybody else. Sometimes, however, vendors can create the most restrictive contracts. In some cases, for example, a manufacturer will offer a maintenance contract requiring that equipment be shipped to the manufacturer and back at customer expense before maintenance can take place. This type of contract is not popular with users. However, many consultants recommend it as a last resort when no outside contractor is available. In many cases, the firms using this kind of maintenance service do not want to commission third-party providers because they foresee a loss in overall maintenance revenues. In reality, however, shipping costs are usually minor when compared to downtime and other losses.

Some original suppliers have been known to be motivated by greed, charging unreasonable fees to what they perceived to be a captive clientele. Some believe they have a built-in monopoly because no one can penetrate their proprietary designs. They can become unresponsive to maintenance calls and complaints, causing a customer to switch to another maintenance contractor.

Total alienation, however, can be avoided. The best way to proceed often is to access the upper channels of the suppliers' management. You don't have to bail out at the beginning—unless your rightful complaints fall on deaf ears. Sometimes faulty business strategies or careless personnel can undermine a vendor's ability to provide service.

Self-Maintenance

The do-it-yourself trend has also hit the imaging world. Some call it "the ultimate form of going bare." However, regardless of how large your company is and how brilliant your in-house staff is, no one uses this strategy for initial maintenance. Your company would be confronted immediately by all of the problems facing the third-party provider, but without being prepared for them. In short, you can't possibly have their background and intricate knowledge of their products (yet).

Once a system is up and running, self-maintenance becomes a more viable option. Still, it's not for everyone. Self-maintenance only makes sense if your organization is sufficiently familiar with the equipment and all of its idiosyncrasies. You also need a supply of spare parts and diagnostic equipment, or at least access to a supply. Finally, you must have the time available to devote to maintenance. Therefore, in general, only large companies are usually in a position to use this strategy.

Best Course of Action

Conventional wisdom says that the first time around, the original supplier provides your maintenance. The original supplier might make a little extra profit, but your investment is safeguarded. If original supplier's services are not readily available, or are simply too expensive, the next best deal is with the third-party providers. Avoid "going it alone."

Many consultants recommend that you keep the maintenance contract with the original supplier as long as the annual costs are not unreasonable. Most of the time, this is the case. The original supplier, unless disaster strikes in one form or another, can usually be the most cost-effective form of maintenance.

What do you do if you use more than one vendor? Multiple brand-name servicing, once the exclusive domain of independents, now is available from about three out of four computer manufacturers. IBM first entered the world of multivendor servicing in 1986. At the time, this was a defensive move to fight penetration made by the independents into IBM accounts. However, IBM, like many other equipment manufacturers, found that servicing other company's products was a profitable business.

You should, however, consider shifting to a third-party supplier when sizable amounts of money can be saved by a reasonably competent outside service. In fact, you would be wise to develop more than one potential third-party supplier. Not only is competition helpful, but you need to develop a contingency plan to maintenance services. Indeed, there are many service vendors out there today willing to give you the exact amount of service and support you want.

Besides, many value-added resellers (VARs) and dealers are turning to partnerships with third-party support organizations to expand their businesses beyond their traditional geographic areas. VARs would rather not rely on affiliates of a national chain or VAR consortia members for remote support for fear of losing the account to its partner. For example, GTE acquired Contel and Telos. Computec acquired R&M Associates, RTK, and part of Computer Maintenance. Diebold and IBM have formed a joint venture to service ATMs. ICL/Fujitsu and Sorbus (both in the U.K.) also have formed joint service ventures.

The trend toward partnering will continue with the advent of open systems. Most vendors realize that they can't be all things to all people. Unlike the past decade, which saw the push toward one-stop service (one vendor handles all support requirements), the coming decade will see a move toward partnering. Indeed, partnering will change the whole delivery methodology of the service industry. OEMs, independents, and dealer channels are going to have to work together as subcontractors to each other in some situations, while remaining competitors in others. While walking the fine line between a sometime-competitor and sometime-partner is not an easy adjustment, it is necessary for survival in the new professional support and open systems environment.

Account control remains an issue at these sites, however. For example, IBM prefers to control all maintenance activity at the site. It wants to be the primary contact, overseeing and administrating all support and service. Under its ideal service partnering arrangement, the customer calls only IBM, which manages the service requirement. If it is not an IBM task, IBM notifies the service partner, oversees its performance, and invoices the customer for the work. In other words, from the customer's standpoint, it still looks like one-vendor service.

Hewlett-Packard, on the other hand, prefers to set up service arrangements allowing the customer to call any of the service partners. The responding company then takes responsibility for determining who should be called and for contacting the proper affiliate service company. Usually, the user maintains separate contracts and invoices with each service vendor.

Software Maintenance

Software maintenance is a tricky topic; even experts disagree on what constitutes software maintenance. Some take a broad approach and define it in terms of anything developers do to maintain the software. Others emphasize the complexities of reengineering and restructuring. The bottom line is that software needs to be adjusted continuously to keep up with changes in business planning.

Maintaining a software program is crucial, but often made difficult by poor planning and documentation. Because system maintenance requires fixing source code written by someone else, documentation needs to be accurate. Without tools that are capable of accurately analyzing and testing the software, the gap between software development and software maintenance can become enormous.

Look at software development as an ongoing process. Software maintenance is part of the project lifecycle, which starts with a request or problem and continues through installation and finally refinement, such as debugging. A typical lifecycle, including maintenance, would look like this:

- Capture the request
- Evaluate current installations
- Test the problem
- Redesign
- Reprogram
- Test
- Implement

Upgrading: Maintenance at a Later Day

Once your system has been in operation for a while, you have more options. This is because, in part, the initial period of risky operation has been completed. Problems caused by manufacturing errors will show right away. In fact, the warranty period should outlast all outright defects in manufacture.

This does not mean that all potential problems have magically disappeared. Unanticipated design weaknesses will often not appear until down the road sometime. These deficiencies can be severe. Inadvertent design deficiencies often cause persistent maintenance problems that need major attention.

The good news is that, in many cases, maintenance eventually becomes predictable. Suppose, for example, you have a typical one-year maintenance contract following a 90-day warranty. That means 15 months will elapse under relatively safe maintenance conditions before you need to consider switching approaches. During this period, most design weaknesses should be revealed. With a 15-month track record behind you, you can reasonably predict malfunctions and downtime. In other words, you have a good idea of the risks involved in maintenance. Reasonable projections allow you to begin comparing alternatives. These alternatives can be evaluated both in terms of cost and risk.

Appendix A

Resources

Publications and Reports

All of the following publications are available from the Association for Information and Image Management (AIIM) resource center located at 1100 Wayne Avenue, Suite 110, Silver Spring, MD 20910.

(ARMA) News, Notes & Quotes
3X/400 Information Management
A-E-C Automation Newsletter
ABA Banking Journal
Access to Wang
ADP Protest Report
Advanced Imaging
Advertising & Marketing News
Advertising Age
AIIM Clips
AIIM Conference Daily
AIIM Job Bank Bulletin
AIIM National Capitol Chapter
AISP Dialogue
AIXpert Journal
ALIC (Archives Library Information Center)
American Archivist
American Library Association Yearbook
American Printer
ANBAR Abstracts: Information Management & Technology

ANSI Reporter
ANSI Standards Action
Annotation
ARMAWORKS
ARMM Blue Line
Architectural & Engineering Systems
Archives and Museum Informatics
Archive & Computer
ASIS Jobline Newsletter
ASIS Proceedings
Aslib Information
Association & Society Manager
Association Management
Association Publishing
Association Trends
At the Centre
Bank Management
Bank New Product News
Bank Systems & Technology
Bank Technology News
Beyond Computing
Bottom Line—Personal
Bottomline Endata
Bull. Bibl. Paris
Bulletin of the American Society for Information Science
Business Advocate
Business Marketing
Business Month
Business Review
Business Week
Byte
CALS Briefing International
CALS Journal
CALS Report
CAP International ODS Newsletter
CAP International Outlook
CAPCON Newsletter
CASE Directions
CASE Trends
Cadence
Candle Computer Report
Capital Computer Digest
Capital PC Monitor
Capital Source

Career Opportunity Update
Catalyst
Cause/Effect
CCC Report
CD Publisher News
CD-I News
CD-ROM Librarian
CD-ROM Professional
CD-ROM Report
Chief Information Officer Journal
CIO Magazine for Information Executives
College Park
Commission on Preservation and Access
Communication World
Communications Briefings
Communications Products & Systems
Communications Standards Management
Communications Standards Review
Communications Week
Computer and Communications Decisions
Computer Buying World
Computer Digest
Computer Graphics Review
Computer Graphics Today
Computer Graphics World
Computer Industry Forecasts
Computer Technology Review
Computer-Aided Engineering
Computers in Banking
Computers in Healthcare
Computerworld
Computerworld Focus
Consensus
Consultant Update
Court Technology Bulletin
Credit Card Management
Data
Data Communications Management
Data Processing Digest
Data Sources
Data Training
Datamation
Datapro Imaging Analyst
Datapro Office Products Bulletin

Datapro Plus (and Significa)
Datapro Reports on Document Imaging Systems
DEC Professional
DEMA
Delibros
Design Graphics World
Design Management
Desktop Publisher
Digital News
Direct
Directory of Online Databases
Disaster Recovery Journal
DODISS Notice
Document Image Automation
Document Image Automation Update
Document Image Processing
Document Management
Document Management Technology
DP Budget
Dr. Dobbs Journal
Eastern Review
EDI News
EDMS (formerly Text Management Journal)
Electric Word
Electronic Imaging Report
Electronic News
Electronic Publishing & Printing
EPSIG News
Exchange (NASIRE)
Executive Update
Federal Computer Week
Financial Imaging News
Financial Technology in the Nineties
Folio
Forbes
Forms and Label Purchasing
Formtek Journal
fyi/im
Geyer's Office Dealer
Governing
Government Computer News
Government Data Systems
Government Technology
Graphic Arts Literature Abstracts

Graphic Arts Monthly
Graphic Arts Product News
Handbook of Optical Memory Systems
Harvard Business Review
HD—High Definition Production
HD World Review
HDTV World Review
High-Tech Marketing
ID Systems
IDC Market Interface
IEEE Standards Bearer
IMC Bulletin
IMC Journal
IMC Proceedings
Image Management Newsletter
Image Management Technology Service
Imaging and Information Consultant
Imaging Business Report
Imaging Magazine
Imaging Quarterly
Imaging Service Bureau News
Imaging Technology Report
ImagingWorld
IN: The Magazine for the Information Age
In-Plant Printer
In-Plant Reproductions
Infocus
Inform
Information Standards Quarterly
Information Broker
Information Center
Information Center Quarterly
Information Executive
Information Industry Bulletin
Information Management & Technology (formerly *Information Media & Technology*)
Information Management Review
Information Science Abstracts
Information Standards Quarterly
Information Standards Quarterly
Information Systems Management (formerly *Journal of Information Systems Management*)
Information Times
Information Times Bulletin

Information Today
Information Week
Infosystems
Infotech
Infoworld
Inside Operations—The Candle View
Insight
Insight (On the News)
Instruction Delivery Systems
Insurance and Technology
Integrated Image
Interact Journal
Interactive World (formerly *4th Media Journal*)
Interleaf Milspec Update
International Journal of Micrographics & Optical Technology
International Micrographics Source Book
International Spectrum
Item Processing Report
Journal of Electronic Imaging
Journal of Engineering, Computing and Applications
Journal of European Business
Journal of Imaging Science and Technology
Journal of the American Society for Information Science
LAN Magazine
LAN Technology
Laser Focus & Electro-optics
Laser Focus World
Lasers and Optronics
LEXIS-NEXIS Insight
Legality of Microfilm
Legality of Micrographics (updates)
Legality of Optical Disk (updates)
LINK-UP
Library Hi Tech News
Library Journal
Library Notes, Naval War College
Library Resources and Technical Services
Library Technology Reports
Log/On
Lotus: Computing for Managers & Professionals
MAC Week
Macazine
Macintosh Buyer's Guide
Macintosh Today

Macintosh-Aided Design
Magazine Design & Production
Magazine Week
Mainframe Journal
Maintenance Technology
Management World
Marketing Computers
Marketing Library Services (MLS) Newsletter
Martian Chronicle
Memoires Optiques
Memory Card Systems & Design
MIS Quarterly
Michigan Business
Micro Forum (ZVtron Newsletter)
Micro Notes
MicroCAD News
Microfilm & Video Systems
Microform Review
Micrographics Market Place
Micrographics Newsletter
Microinfo Microfilm News
Modern Office Technology
Multimedia & Text Report
Multimedia & Videodisc Monitor
NARA Bulletin
NASA Tech Brief
Nation's Business
National Journal
Network Computing
Network SEI
New York Times
News & Review
News From The Archives
Newsweek
NIRMA Newsletter
NIRMA Symposium News
NOMDA Spokesman
NOPA Industry Report
NSIA News
NTIS Alert (Formerly *NTIS Administration & Management*)
OE Reports
OE&M—Office Equipment & Methods
Off the Shelf
Office

Office Market Update
Office Systems
Office Systems and Technology
Office Systems Management
Office Technology Management (formerly *Today's Office*)
Office World News
OMNI
Omnicom Information Services
Optical & Magnetic Report
Optical Engineering Reports
Optical Information Systems
Optical Memory News
Optics in Education
Panorama
PC Magazine
PC Publishing
PC Week
PC World
Peopleware
Personal Computing
Personal Engineering and Instrumentation News
Personal Report
Photo Marketing
Plan and Print
PMA Newsletter
Popular Science
Publish!
Publishing & Production Executive
RDBMcomputing Magazine
Records & Retrieval Report
Records Management Quarterly
Regardie's
Repro Report
Rythms
S. Klein Newsletter on Computer Graphics
SAA Newsletter
Scanning Journal
Scientific Computing & Automation
Scientist
Scoop Sheet
Selling Space
Service Bureau Newsletter
SIGCAT Newsletter
SIGOMET Newsletter

SIM Spectrum
Society for Applied Learning Technology Newsletter
Software Magazine
Software News
Solutions
Soundview Executive Book Summaries
Standardization Newsletter (was *Standardization & Data Management Newsletter*)
Standardization Directory
Standardization News
Standards Engineering
Status of Standardization Projects—SD4
Success
Successful Magazine Publishing
Systems Integration
T & E News
T.H.E. Journal
TBS Newsviews
Tech Exec
Technical Leaflet
Technical Photography
Technology Review
Telecommunication Products Plus Technology
Telecommunications
Telemarketing
Time
True Imaging
U.S./R&D
U.S.Dept. of Commerce News
UNISYS Systems
Unix/World
Update
USA Today
USAE (U.S. Association Executive)
Varbusiness
Videodisc and Optical Disk
Videodisc Monitor
Videography
Wall Street Journal
Wang News
Washington Business Journal
Washington Business Review
Washington Post
Washington Report

Washington Technology
Western Banker
Where
Words

Details on Selected Publications

BYTE
McGraw-Hill, Inc.
One Phoenix Mill Lane
Peterborough, NH 03458
603-924-9281
Comprehensive computer magazine.

CD Data Report
Helgerson Associates, Inc.
510 North Washington Street, Suite 401
Falls Church, VA 22046-3537
703-237-0682

Computer Systems News
CMP Publications, Inc.
600 Community Drive
Manhassett, NY 11030
516-562-5000
Newsweekly for the computer and systems integration business.

Connect (The Journal of Computer Networking)
3Com Corporation
3165 Kifer Road
Santa Clara, CA 95052-8145
Aimed primarily at information systems and data processing managers, this free quarterly magazine discusses issues of interest to all LAN users, not just users of 3Com products.

Data Communications
McGraw-Hill Information Services
McGraw-Hill Building
1221 Avenue of the Americas
New York, NY 10020
Phone: 212-512-2000
Fax: 212-512-6833
Considered by many to be the journal of record in the communications and computing industries, this monthly magazine covers both technical and marketing issues.

Data Entry Awareness Report
Management Information Corporation
401 East 70
Cherry Hill, NJ 08034
609-428-1020
Provides evaluations of such data entry products as intelligent terminals, character readers, portable scat recorders, direct data entry, videotex for banks, retail and factories.

Data Entry Management Association Newsletter
Data Entry Management Association
101 Merritt 7 Corporate Park, 5th Floor
Norwalk, CT 06856
203-846-3777
Provides articles on data capture and management, including equipment, statistical compensation survey, and video terminal health survey.

Data Processing Digest
c/o Gisela I. Wermke
Box 1249
Los Angeles, CA 90078-1249
213-851-3156
A digest of articles of interest to data processing professionals and information managers.

Data Processing Management Association Magazine
c/o Amdahl Canada
1 First Place
P.O. Box 123
Toronto, Ontario M5X
Tabloid covering minimicrosystems.

Data Training
38 Chauncy Street
Boston, MA 02111
617-542-0146
Features training in information maintenance, office automation, and information training.

Electronic Imaging Report
Electronic Messaging News
Electronic Trade And Transport News
Phillips Publishing, Inc.
7811 Montrose Road
Potomac, MD 20854
301-340-2100

These three publications, from the same publisher, cover aspects of the electronic information industry.

FYI/IM
Association for Information and Image Management
1100 Wayne Avenue
Silver Spring, MD 20910
301-587-8202
Newsletter covering AIIM news and industry developments.

Home Office Computing
Scholastic Inc.
740 Broadway
New York, NY 10003
Editorial: 212-505-3580
Subscription: 800-678-0118
Includes reviews of hardware and software from the home office viewpoint and profiles of successful home businesses.

Inc. Magazine
Goldhirsh Group
38 Commercial Wharf
Boston, MA 02110
Editorial: 617-248-8000
Subscription: 800-234-0999
Written for the executives of small and mid-sized companies.

Imaging
The Imaging Industry Magazines
12 West 21st Street
New York, NY 10012
212-691-8215
Magazine highlighting all aspects of imaging.

Information Week
CMP Publications, Inc.
600 Community Drive
Manhassett, NY 11030
Phone: 516-562-5000
Fax: 516-562-5474
This weekly business publication focuses on the entire information industry, primarily for the benefit of information managers. However, its information is also suitable for individuals with considerably less expertise.

Inform: The Magazine Of Information and Image Management
Association for Information and Image Management
1100 Wayne Avenue
Silver Spring, MD 20910
Magazine featuring in-depth discussions of current trends in the electronic imaging field as well as spot news coverage.

InfoWorld
1060 Marsh Road
Menlo Park, CA 94025
800-344-4636 (800-227-3265 in California)
This weekly trade paper is noted for its in-depth reviews of LAN and PC products.

Microcomputer Applications
ISMM Acta Press
2990 Red Hill Road
Costa Mesa, CA 92626
Covers engineering, science, business, management and other microcomputer applications.

Microcomputer Communications
Faulkner Technical Reports
114 Cooper Center
7905 Browning Road
Pennsauken, NJ 08109
Product and management reports covering microcomputer communications technologies, including linking services.

PC Magazine
1 Park Avenue
New York, NY 10016
212-503-3500
Published twice monthly, this magazine offers up-to-the-minute coverage of the PC marketplace. It features articles on new products, product reviews, and several continuing advice columns.

PC Week
800 Boylston Street
Boston, MA 02199
615-375-4000
This weekly publication features several special insert sections, such as the one called "Connect," which deals primarily with networking.

The Records & Retrieval Report:
The Newsletter For Professional Information Managers
Greenwood Publishing Group
88 Post Road West
Westport, CT 06881
203-226-3571
This newsletter provides guidelines for professionals interested in cost-efficient records management.

Books

The AIIM resource center includes a wide selection of books. In addition, AIIM has its own bookstore catalog. You can order books or a catalog by calling 301-587-8202. The following are a few selections.

Imaging

DataPro Reports on Document Imaging Systems
Datapro Research
600 Delran Parkway
Delran, NJ 08075
800-328-2776
This 1,000-page report covers the technologies, markets, and vendors of imaging systems and includes a comprehensive list of vendors.

Imaging: Concepts and Implementation
Seybold Office Computing Group
148 State Street
Boston, MA 02109
617-742-5200
This report provides a detailed description of imaging technology, vendors, and relevant applications and contains numerous helpful case studies.

Image Processing: The Next Frontier
The Yankee Group
200 Portland Street
Boston, MA 02114
617-367-1000
A report that covers the industry, market potential, suppliers, users, and applications, as well as the general state of imaging technology.

Implementing Electronic Imaging: A Management Perspective
By Scott Wallace
Cutter Information Corporation
37 Broadway
Arlington, MA 02174
617-648-8700

As the title indicates, this well-received report caters primarily to the needs of managers, providing a detailed roadmap for the process of implementing electronic imaging.

Introduction to Electronic Imaging
By Don M. Avedon
Association for Information and Image Management
1100 Wayne Avenue
Silver Spring, MD 20910-5699
This book tells how electronic image management (EIM) can help any organization stymied by paper overload. It covers the advantages of EIM systems, available media, standards and legality issues, system integration and workflow, system evaluation, the art of sizing and scoping a document, indexing, backfile conversion, and applications analysis.

Management

Proposal Development: How to Respond and Win the Bid
By Bud Porter-Roth
The Oasis Press
300 North Valley Drive
Grants Pass, OR 97526
Whether you are bidding on a $5,000 project or a $10 million contract, the way your proposal is written and presented can mean the difference between winning and losing. This book offers a practical step-by-step guide to the mechanics of putting together a proposal.

Strategic and Operational Planning for Information Services
QED Information Sciences
170 Linden Street
Wellesley, MA 02181
This how-to guide for short and long-range planning shows you how to plan, how to apply plans for specific decision situations, and how to avoid common errors and difficulties. Emphasis is on integrating MIS systems, planning procedures, and policies into larger organizations, as well as organizing MIS planning functions themselves.

Organizations

The most important organization for anyone interested in imaging systems is the Association for Information and Image Management (AIIM).
1100 Wayne Street
Silver Spring, MD 20910-5699
301-587-8202

For information about technical support groups for specific hardware or software or for using computers in specific industries, refer to *The Encyclopedia*

of Associations. For more information on the Encyclopedia of Associations, contact Gale Research
835 Penobscot Building
Detroit, MI 48226

College Courses

Bowling Green State University
Department of Business Education
Bowling Green, OH 43403
419-372-2747

- Office and Records Management Systems In this course, a discussion of micrographics is included with other aspects of record management.

Catholic University of America School of Library and Information Science
Washington, DC 20064
202-319-5085

- Introduction to Information Resources Management This one-semester course is an introduction to information as an organizational resource. Information resources management within its social and cultural context is discussed. The course covers a wide range of information resource management problems including information economics, information decay, information recycling, alternate approaches to measuring the value of information, information organization structures, and the role of information managers.
- Information Storage and Retrieval Systems A one-semester course that covers analysis and management of records from creation through processing, maintenance, protection, and final disposition. The course examines the components of integrated records management programs, the major areas of records management, the tools and techniques needed for managing records, and the impact of records management systems on various organizations.
- Information Sources and Services A one-semester course that examines the information resource environment of large organizations to determine how principles of information resources management may be applied.

Colorado State University
Division of Continuing Education
Fort Collins, CO 80523
800/525-4950 or 303/491-8417

- Telecommunications Extended Studies Videotaped courses provide usable knowledge to meet training and development needs.

Columbia Union College
7600 Flower Avenue
Takoma Park, MD 20912
301-891-4073

- Records Management A one-semester course in the methods and systems of storing and retrieving information. Provides practice in filing and information on organizing a records program and developing effective operating procedures and controls. Course also discusses personnel qualifications.

El Centro College
The Business Division
Main and Lamar
Dallas, TX 75202

- Records Control (OFC 250) This three-lecture course includes a comprehensive study and application of the knowledge and skills involved in the control of records and record systems. The course examines control procedures for the management of routing and unique correspondence, directives, proposals, reports and forms, inventory, scheduling, vital records control, records storage centers, and archives.
- Introduction to Records Management (OFC 152) A three-lecture survey course in the policies and principles affecting the creation, protection, circulation, retrieval, preservation, and control of business and institutional records. The course includes basic classification systems, the history and status of records management, the retention and disposition of records, maintenance procedures, and career ladders.
- Filing Practices (OFC 150) Composed of two lectures and two lab sessions, this course introduces basic principles and procedures of records storage and control. Topics include records storage methods, procedures for the operation and control of manual and automated storage systems, rules for indexing, and principles for the selection of records equipment and supplies.

George Washington University Engineering and Applied Science
Continuing Engineering Education Program
Washington, DC 20052
800-424-9773 or 202-994-6106

- Micrographics and Optical Disk Systems A two- to three-day course. Contact the continuing engineering education program for details on content.
- Systems Analysis Techniques for Information Managers (902DC) A two-day course in which participants learn to identify and define information system requirements, evaluate alternatives, assess costs, and prepare effective reports on study results. The course presents systems analysis methodologies in broad contexts, applicable to manual and automated information-handling problems. Beneficial to information managers, librarians, and information processing professionals in industry, government, and business.
- Videodisk Technology and Applications (1 154DC) A two-day course that introduces videodisk technology and its applications for information stor-

age and retrieval and to help participants apply videodisk technology to the rapidly expanding information revolution in business and government.

- Optical Disks: Technologies and Applications (1269DC) This two-day course provides an overview of the information storage media that are becoming commercially available, and evaluates and compares them in terms of technology, applications, markets, and user economics. On-site seminars are conducted on all aspects of information and imaging technology.

Howard Community College
Business Management Division
Columbia, MD 21044
301-992-48927

- Records Management A one-semester course that studies records storage systems and procedures for controlling office records.
- Modern Office Procedures This one-semester course includes problem solving, time management, listening skills, and work prioritization. Course also discusses taking minutes, planning travel arrangements, and telecommunications procedures.

Mankato State University
Business Education Department
Mankato, MN 56001
507-389-6116

- Information And Records Management (BE 201) The objective of this one-quarter course is to become acquainted with current developments in data processing that relate to information storage and retrieval.

Montgomery College
Takoma Park Campus
Takoma Avenue and Fenton Streets
Takoma Park, MD 20912
301-587-4090

- Filing Systems and Records Management One semester (two hours per week) that serves as an introduction to the filing systems and records management requirements of business, including indexing, coding, and cross-referencing principles, filing techniques, and basic principles of retention. The student will practice identification, storage, and retrieval methods.

Rutgers University
School of Communication
Information and Library Studies
4 Huntington Street
New Brunswick, NJ 08903
201-932-7500

- Information and Communications Technology (17610523) A six-week course that examines noncomputer technologies used in communication and information services (micrographics, lasers, optics, holography, electronic printing, reprography, etc.).

San Diego State University
College of Extended Studies
San Diego, CA 92182
619-594-5821

- Introduction to Telecommunications and Telecommunications Planning And Management A course about the use of microforms in information retrieval systems.

State University of New York at Albany
School of Information Science & Policy
135 Western Avenue
Albany, NY 12222
518-442-5115

- Records Management Including ODDD, CD-ROM, and WORM (ISP646) A one-semester course that covers basic concepts and practices of records management in governmental, institutional, and corporate agencies, including those areas of communication, administration, and computer technology that relate to the efficient and effective flow of information from its generation to its final disposition. Course includes records inventory, active and inactive records control, manual and automated systems, vital records protections, the records center, and legal and ethical aspects of records management.
- Information Systems and Technical Applications (ISP61 1) Course discusses technical applications including computing, telecommunicating, and formatting. Also covers reprography, the design of facsimile transmission systems, problems of reprography and copyright, and optical publishing with CD-ROM.

University of the District of Columbia Business Education
Information Processing Technology
43400 Connecticut Avenue
Washington, DC 20008
202-282-3727

- Records Systems Management This one-semester course covers automated and micrographic principles governing management of business office records, techniques for managing private and government records, and long-term records issues. Includes practical experience in a laboratory setting.

University of Maryland
College of Library and Information Sciences
College Park, MD 20784
301-454-3016

- Principles of Records Management and Machine Readable Files This one-semester course provides an introduction to practices and terminology of records management within the broad framework of information management.
- Seminar on Information Resources Management (IRM) Content for this one-semester course is drawn from many disciplines in order to plan and manage information systems, services, and technologies within organizations. These disciplines include information science, computer science, and business administration. This course reviews the success and problems of IRM since 1980, with emphasis on tools to manage information resources.

University of North Dakota
Department of Business and Vocational Education
Grand Forks, ND 58201
701-777-2517

- Records Management I and Records Management II There is a substantial micrographics introduction in Records Management I and a larger micrographics systems segment in the Records Management II course. Courses include discussion of COM and optical disk systems.

University of Wisconsin-Eau Claire
Department of Business Education and Administrative Management
Eau Claire, WI 54702-4004
715-836-4320

- Records Management (316/516) This course covers controlling costs and improving the effectiveness of records and information management within business organizations. The course discusses the control of record creations, maintenance, and disposition through systems analysis, classification systems, storage locations and equipment, protection methods, legal retention policies, and forms management.
- Document-Based Management Systems Microfilm as an alternative to paper and electronic media for the management of document-based information is covered. The course includes source document microfilm, computer-output microfilm, microfilm feasibility, and systems design. The emphasis is on computer-assisted document storage and retrieval systems, and emerging electronic technologies for the management of documents.

University of Wisconsin-Whitewater
800 West Main Street
Whitewater, WI 53194
414-472-1234

- Records Management (220-445/645) This course covers the development of records administration principles including creation, use, maintenance, and destruction. A discussion of storage facilities, records, classification, forms and reports control, vital records protection, and micro-image systems is also included.

U.S. Government
Graduate School USDA
600 Maryland Avenue, SW Room 115-116
Washington, DC 20024
202-447-7124

- Technical Excellence: Micrographics Management Techniques (SINFO745) One-day workshops aimed at persons working in the fields of information resources, policy and procedures, records management, and Micrographics. These workshops provide information and training in quality assurance, archival storage, and equipment troubleshooting.

National Archives and Records Administration Agency
Services Division (NIA)
Washington, DC 20408
202-501-6025

- Offers courses in Introduction to Records Management, Files Improvement, Records Dispositions, Scheduling Records, Managing Audiovisual Records, Managing Electronic Records, and Managing Record Dispositions.

Industry Contact List

A good way to find out what is going on in any industry is to research the key players. In the electronic imaging industry, those key players usually make an annual pilgrimage to the AIIM trade show held every spring. The following list, compiled from the 1992 and 1993 registrants, might be the closest the industry has to a readily available contact list.

Show Exhibitors for the 1992/1993 AIIM Shows included the following:

Adaptive Information Systems
24461 Ridge Route Drive
Suite 200
Laguna Hills, CA 92653
800-678-6978

Advanced Technology Systems
800 Follin Lane, Suite 270
Vienna, VA 22182
703-242-0300

AEG
1350 Connecticut Avenue NW
Suite 410
Washington, DC 20036
202-835-2003

Alliance Data Systems
555 Twin Dolphin Drive, Suite 240
Redwood City, CA 95065
415-591-2000

Alliance INFONET
4675 MacArthur Court, Suite 600
Newport Beach, CA 92660
714-863-6900

ALOS Micrographics
18 Bracken Road
Montgomery, NY 12549
914-457-4400

Alpharel
3601 Calle Tecate
Camarillo, CA 93012
805-482-9815

Amitech
2721-E Merrilee Drive
Fairfax, VA 22039
703-698-5057

Apogee Computer Systems
6825 East Tennessee Avenue
Suite 500
Denver, CO 80224
303-320-6378

Applied Programming Technologies
One Hollow Lane, Suite 313
Lake Success, NY 11042
516-365-1098

Aquidneck Systems International
650 Ten Rod Road
North Kingstown, RI 02852
401-295-2691

ATLIS Image Systems
6011 Executive Boulevard
Rockville, MD 20852
301-770-3000

Avision
Science-Based Industry Park
2nd Floor, #2
Prosperity Road II
Hsinchu, R.O.C. Taiwan
886-35-782-388

Bell & Howell
6800 McCormick Road
Chicago, IL 60645
708-675-7600

Benson Computer Research
7926 Jones Branch Drive, Suite 260
McLean, VA 22102
703-442-4545

Blueridge Technologies
Flint Hill Square
Flint Hill, VA 22627
703-675-3015

Boss Logic
505 North 3rd Street
Fairfield, TX 52556
515-472-7740

Borett Automation Technologies
31324 Via Colinas, Suite 106
Westlake Village, CA 91362
818-597-8664

BULL HN Information Services
Technology Park
Billerica, MA 01821
508-294-4534

CalComp Digitizers Product Group
14555 North 82 Street
Scottsdale, AZ 85260
602-948-6540

Calera Recognition Systems
475 Portero Avenue
Sunnyvale, CA 94086
408-720-1330

Candi Technology
2236 Camino Ramon
San Ramon, CA 94583
510-866-1226

Canon USA
One Canon Plaza
Lake Success, NY 11042-1113
516-488-6700

Capital Data
3300 West Main Street
Lansing, MI 48917
800-999-4409

CCR (Combined Computer Resources)
5429 LBJ Freeway
Lock Box 110
Dallas, TX 75240
214-851-7773

Cimage
3885 Research Park Drive
Ann Arbor, MI 48108
313-761-6550

CISD
3525 Hyland Avenue, Suite 260
Costa Mesa, CA 92626
714-641-4300

Coastal Software
100 Commercial Street
Portland, ME 04101
207-871-9366

Cognitronics Imaging Systems
4780 Mission George Place
San Diego, CA 92120
619-265-3434

Com Com Systems
2420 Enterprise Road, Suite 201
Clearwater, FL 34623
813-725-3200

Computron Technologies
301 Route 17 N.
Rutherford, NJ 07070
201-935-3400

COM Squared Systems
1285 Corporate Center Drive
St. Paul, MN 55121
612-452-9522

Congruent
110 Greene Street
New York, NY 10012
212-431-5100

Continental Graphics
163 South LaBrea Avenue
Los Angeles, CA 90036
213-938-2511

Cordent Imaging Systems
50 Stiles Pond Road
PO Box 462
West Boxford, MA 01885
508-352-5500

The Core
PO Box 50845
Pasadena, CA 91115
818-796-9155

Corel System
1600 Carling Avenue
Ottawa, Ontario KIZ 8R7
613-728-8200

Cornerstone Technology
1990 Concourse Drive
San Jose, CA 95131
408-435-8940

Corporate Information Systems
29 16 Chicago Drive SW
Grandville, MI 49418
616-538-5550

CPT Image Systems
8100 Mitchell Road
Eden Prairie, MN 55344
612-937-8000

Cranel
51OG East Wilson Bridge Road
Worthington, OH 43085
614-433-0045

CSI/CAD Scan
2200 Menaul Boulevard NE, Suite B
Albuquerque, NM 87107
505-881-4872

CTA
25 Science Park
New Haven, CT 06511
203-786-5828

Cuadra Associates
11835 West Olympic Boulevard
Suite 855
Los Angeles, CA 90064
310-478-0066

Datacap
580 White Plains Road
Tarrytown, NY 10591
914-332-7515

Data Image
628 Hebron Avenue
Glastonbury, CT 06033
203-659-3980

Data Management Services
PO Box 619285
Dallas/Ft. Worth, TX 75261
817-963-3068

Dataworks
45 NE Loop 410, Suite 875
San Antonio, TX 78216-5831
512-344-7494

Dickens Data Systems
1175 Northmeadow Parkway
Suite 150
Reswell, GA 30076
404-475-8860

Digital Equipment Corporation (DEC)
Continental Boulevard
Merrimack, NH 03054
603-884-0157

Diversified Management International
165 N. Plains Indian Road
Wallingford, CT 06492
203-284-0818

DocuFile
462 Stevens Avenue, Suite 102
Solana Beach, CA 92075-2065
619-793-1200

DocuMaster Systems
1420 N. Claremont Boulevard
Building 103-B
Claremont, CA 91711
714-625-0017

Document Technologies
1300 Charleston Road
Mountain View, CA 94043
415-964-6100

DocuPoint
2701 Bayview Drive
Fremont, CA 94538
510-770-1189

DSK Technology
671 Via Alondra, Suite 808
Camarillo, CA 93012
805-484-3696

Elke Microforms
3352 Malher Field Road
Rancho Cordova, CA 95670
916-368-5582

Epoch Systems
Eight Technology Drive
Westborough, MA 01581
508-836-4300

Excalibur Technologies
2000 Corporate Ridge, Suite 105
McLean, VA 22102
703-790-2110

Executive Technologies
2120 16 Avenue South
Birmingham, AL 35205
205-933-5494

ExperVision
3590 N. First Street
San Jose, CA 95134
408-428-9444

Felth Systems & Software
425 Maryland Drive
Fort Washington, PA 19034
215-646-8000

FileNet
3565 Harbor Boulevard
Costa Mesa, CA 92626
714-966-3400

Fujitsu Computer Products of America
2904 Orchard Parkway
San Jose, CA 95134
1-800-626-4686

Fulcrum Technologies
785 Carling Avenue
Ottawa, Ontario KlS 5H4
613-238-1761

GeneSys Data Technologies
Four North Park Drive, Suite 400
Hunt Valley, MD 21030
410-785-0660

GESCAN
PO Box 12599
Research Triangle Park, NC 27709
919-460-3100

Greengage Development
1895 Park Avenue
San Jose, CA 95126
408-243-8980

GTE Vantage Solutions
15000 Conference Center Drive
Chantilly, VA 22021-3808
703-818-5742

Hewlett-Packard
19091 Pruneridge Avenue
Cupertino, CA 95014
408-447-1378

HNC
5501 Oberlin Drive
San Diego, CA 92121
619-546-8877

Horizons Technology
3990 Ruffman Road
San Diego, CA 92123
619-292-8320

IBM
1133 Westchester Avenue
White Plains, NY 10604
914-642-6037

IBS
417 Fifth Avenue, 5th Floor
New York, NY 10016
212-696-2500

ICT Imagedata
Bessemer Road
Welwyn Garden City
PO Box 6
Shire Park Herts A17 1HD England
44-707-337284

Ideal Scanners
11810 Parklawn Drive
Rockville, MD 20852
301-468-0123

IdentiTech
1333 Gateway Drive
Melbourne, FL 32901
407-951-9503

ILevy Associates
1633 Des Peres Road, Suite 205
St. Louis, MO 63131
314-822-0810

Image Data
11550 IH-10 West
San Antonio, TX 78230
512-641-8340

Image Machines
590 Herndon Parkway
Herndon, VA 22070
703-709-7475

Image Network Technology
2550 Corporate Place, Suite C101
Monterey Park, CA 91754
213-263-9153

Image Processing Technologies
8300 Boone Boulevard, Suite 500
Vienna, VA 22182
703-848-9226

Image Systems
Rensselaer Technology Park
Troy, NY 12180
518-283-8783

Image Tech
29444 Northwestern Highway
Suite L-500
Southfield, MI 48034
313-353-7900

Image-X International
600 Ward Drive, B-1
Santa Barbara, CA 93111
804-964-3535

Imars Research
111 Peter Street, Suite 804
Toronto, Ontario M5V 2HI
416-581-1740

Indus International
340 South Oak Street
West Salem, Wl 54669
608-786-0300

Information Management Consultants
7915 Westpark Drive
McLean, VA 22102
703-893-3100

Ingram Micro
1600 Saint Andrew Place
Santa Ana, CA 92795
714-566-1000

Integraph Technical Information Management
Huntsville, AL 35894
205-730-8302

Integris
900 Middlesex Turnpike
Building Two
Billerica, MA 01821
508-294-7900

Intelus
9210 Corporate Boulevard
Rockville, MD 20852
301-984-6363

Intrafed
5185 MacArthur Boulevard NW
Suite 102
Washington, DC 20016
202-686-2360

Jodian Systems
22648 Glenn Drive, Suite 205
Sterling, VA 22170
703-742-7795

JVC Information Products
19900 Beach Boulevard, Suite 1
Huntington Beach, CA 92648
714-965-2610

Keyfile
22 Cotton Road
Nashua, NH 03063-4219
603-883-3800

Kofax Image Products
3 Jenner Street
Irvine, CA 92718
714-727-1733

Lanier Worldwide
2300 Parklake Drive NE
Atlanta, GA 30345
404-496-9500

LaserCard Systems
2644 Bayshore Parkway
Mountain View, CA 94043
415-969-4428

LaserCOM
2916 Chicago Drive SW
Grandville, MI 49418
616-538-5550

LaserData
300 Vesper Park
Tyngsboro, MA 01879
508-649-4600

Laser Magnetic Storage International
4425 ArrowsWest Drive
Colorado Springs, CO 80907
719-593-7900

Locar Imaging Systems
25 Burlington Mall Road, Suite 300
Burlington, MA 01803
617-270-0665

Logical Software Solutions
7701 Greenbelt Road, Suite 207
Greenbelt, MD 20770
301-474-0285

Logicon Ultrasystems
525 South Douglas Street
El Segundo, CA 90245
310-643-5111

Lotus Development
55 Cambridge Parkway
Cambridge, MA 02142
617-577-8500

Management Data
3300 Cababa Road, Suite 201
Birmingham, AL 35223
205-871-5844

Maxoptlx
2520 Junction Avenue
San Jose, CA 95134
408-954-9700

Metafile Information Systems
421 First Avenue SW
Rochester, MN 55902
1-800-638-2445

Micro Design International
6985 University Boulevard
Winter Park, FL 32792
407-677-8333

Micro Dynamics
8555 16th Street, Suite 701
Silver Spring, MD 20910
301-589-6300

Micrographic Technology (MTC)
520 Logue Avenue
Mountain View, CA 94043
415-962-7460

Microseal
2000 Lewis Avenue
Zion, IL 60099
708-872-1666

3M Office Document Systems Division
3M Center
St. Paul, MN 55144
612-736-7646

Minolta
101 Williams Drive
Ramsey, NJ 07446
201-825-4000

Mira
9861 Copper Hill Road
Saint Louis, MO 63124
314-965-1111

MultiProcess Computer
One Industrial Drive
Windham, NH 03087
603-893-9090

National Computer Systems
4401 West 76th Street
Edina, MN 55435
612-830-7962

Neotech Systems
103 West 61 Street
Westmont, IL 60559
708-963-1202

Nestor
One Richmond Square
Providence, RI 02906
401-331-9640

Network Computing Devices
350 North Bernardo Avenue
Mountain View, CA 94043
415-691-2630

Nikon Electronic Imaging
1300 Walt Whitman Road
Melville, NY 11747
516-547-4355

Northern Development Group
7100 SW Hampton Street, Suite 207
Portland, OR 97223
503-620-0196

Novacad
129 Middlesex Turnpike
Burlington, MA 01803
617-221-0300

OCR Systems
800 Bybeffy Road, Suite 1405
Huntingdon Valley, PA 19006
215-938-7460

OCRON
3350 Scott Boulevard, Building 36
Santa Clara, CA 95054
408-980-8900

Optibase
7800 Deering Avenue
Canoga Park, CA 91304
818-719-6566

Optical Magnetic Imaging
6917 Woodley Avenue
Van Nuys, CA 94118
818-908-0475

Optika Imaging Systems
980 Enchanted Way
Simi Valley, CA 93065
805-520-9060

Optowand
3200 West Story Road
PO Box 166558
Irving, TX 75016-6558
214-257-2100

Pamet Systems
1000 Main Street
Acton, MA 01720
508-263-2060

Panasonic Communications & Systems
Two Panasonic Way 7G-1
Secaucus, NJ 07094
201-348-7000

PaperClip Software
One University Plaza, Suite 309
Hackensack, NJ 07601
201-487-3503

Paperless
1750 North Colhm, Suite 104
Richardson, TX 75080
214-235-4008

Personal Library Sortware
2400 Research Boulevard
Suite 350
Rockville, MD 20850
301-990-1155

Photomatrix
5700 Buckingham Parkway
Culver City, CA 90230
213-417-3800

PI Technology
871 Tapo Street
Simi Valley, CA 93063
805-582-0775

Pinnacle Micro
19 Technology
Irvine, CA 92718
800-553-7070

Pixel Translations
10062 Miller Avenue, Suite 205
Cupertino, CA 95014
408-865-6600

Plasmon Data Systems
1654 Centre Pointe Drive
Milpitas, CA 95035
408-956-9400

Plexus Software
PO Box 660204
Dallas, TX 75266-0204
214-579-6000

QStar Technologies
600 East Jefferson Street
Rockville, MD 20852
301-762-9800

R Squared
11211 East Arapaho Road, Suite 200
Englewood, CO 80112
303-799-9292

Recognita
1156 Aster Avenue, Suite F
Sunnyvale, CA 94086
408-241-5772

Ricoh Peripheral Products
3001 Orchard Parkway
San Jose, CA 95134
408-432-8800

SAIC Imaging Solutions
10770 Wateridge Circle
San Diego, CA 92121
800-442-7242

Satellite Image Systems
990 West Artherton Drive, Suite 202
Salt Lake City, UT 84123
801-262-1792

ScanOptics
22 Presage Park
East Hartford, CT 06108
203-289-6001

SearchExpress
2120 16 Avenue
South Birmingham, AL 35205
205-933-5494

Semaphore Image
400 Lincoln Village, Suite 2331
Larkspur, CA 94939-2206
415-461-1890

Sequoia Data
433 Airport Boulevard, Suite 414
Burlingtowne, CA 94010
415-696-8750

Sletec Open Systems
2235 Sheppard Avenue E, Suite 1800
Willowdale, Ontario M2J 5B5
416-496-8510

Sigma Imaging Systems
622 Third Avenue
New York, NY 10017
212-476-3000

Soda Creek Technologies
72 Via Floreado
Orinda, CA 94563
510-254-1461

Software Alliance
2150 Shattuck Avenue
Berkeley, CA 94704
510-548-7752

SoftwareFABRIK
Phorugasse 8 A-1040
Wien, Austria
43-1-587-689110
U.S.: 714-494-8165

Solea Systems
25251 Paseo de Alicia, Suite 200
Laguna Hills, CA 92653
714-768-7736

Sony Corporation of America
3 Paragon Drive
Montvale, NJ 07645
201-930-6153

Southern Computing Systems
2732 7th Avenue South
Birmingham, AL 35233
205-251-2985

Southern Electric International
100 Ashford Center N., Suite 400
Atlanta, GA 30328
404-261-4700

Syntax
1400 Commerce Park Drive
Suite 220
Reston, VA 22091
703-671-2676

System 2100
8 Woodside Road
Amersham Bucks, HP6 6PA England
494-772-2126

TASC
55 Walkers Brook Drive
Reading, MA 01867
617-942-2000

TEAMWorks Technologies
65 Boston Post Road West
Marlboro, MA 01752
508-460-0053

Telephoto Communications
11722-D Sorrento Valley Road
San Diego, CA 92121
619-452-0903

Terminal Data (TDC)
5898 Condor Drive
Moorpark, CA 93021-2601
805-529-1500

TextWare
PO Box 3267
Park City, UT 84060
801-645-9600

TMS
PO Box 1358
Stillwater, OK 74076
405-377-0880

TRW Financial Systems
2001 Center Street
Berkeley, CA 94704
415-704-3000

Unisys
PO Box 500
Blue Bell, PA 19424-0001
215-986-2243

United Systems and Sortware
2301 Lucien Way, Suite 360
Maitland, FL 32751
407-875-2120

Vantage Computer Systems
100 Great Meadow Road
Welhersfield, CT 06109
203-721-0694

Verity
1550 Plymouth Street
Mountain View, CA 94043
415-960-7600

ViewStar
5820 Sheilmound Street
Meryville, CA 94608
510-652-7827

VisionShape
1434 West Taft Avenue
Orange, CA 92665
714-282-2668

Vividata
2020 Milvia, Suite 414
Berkeley, CA 94704
510-841-6400

Wang Laboratories
1 Industrial Avenue
Lowell, MA 01851
508-459-5000

Appendix

B

Basic Units of Measure

Understanding basic units of measure provides the framework to understanding information in articles, publications, vendor data sheets, etc. Standardized units of measure make sure you are comparing "apples with apples."

A bit (or binary digit) is the smallest unit of digital information. It represents one of two possible discrete values or conditions, usually thought of as a numeric value for either zero or one, on or off, mark or space, something or nothing.

Byte = 8 bits

Thousand bytes = kilobyte (KByte or K) = $10^3 = 2^{10}$ = 1,000 bytes

Million bytes = megabytes (M) = $10^6 = 2^{20}$ = 1,000,000 bytes

Billion bytes = gigabytes (G) = $10^9 = 2^{30}$ = 1,000,000,000 bytes

Trillion bytes = terabytes (T) = $10^{12} = 2^{40}$ = 1,000,000,000,000 bytes

Quadrillion bytes = petabytes (P) = $10^{15} = 2^{50}$ = 1,000,000,000,000,000 bytes

I have given these units of measure to the base 10 and base 2. Most people are accustomed to using the base 10 (as in our monetary system), but computers use a binary system (zero and one), which is base 2. The base 2 orientation of computers makes these numbers slightly larger, i.e., one megabyte is approximately one million bytes but it is precisely 1,048,576 bytes.

Another related area of confusion is bits and bytes, particularly when the terms are abbreviated. For example, when discussing megabits per second the correct abbreviation is *Mbps* (lowercase *b*) and when talking about megabytes per second the correct abbreviation is *MBps* (uppercase *B*). Remember eight bits equal one byte.

Appendix

C

Standards

AIIM Standards

MS44 (ANSI/AIIM MS44-1988)
Recommended Practice for Quality Control of Image Scanners

MS49 (ANSI/AIIM MS49)
Recommended Practice for Scanning from Roll Film and Microfiche

MS50 (ANSI/AIIM MS50)
Recommended Practice for Monitoring Image Quality of Aperture Card Film Scanners (*in development*)

AIIM Technical Reports

TR2 (AIIM TR2)
Glossary of Imaging Technology

TR14 (AIIM TR14-1988 with 1991 addendum)
Interchange of Tiled Raster Document

TR15 (AIIM TR15)
Electronic Image Capture of Documents (*in development*)

TR17 (AIIM TR17-1989)
Facsimile and Its Role in Electronic Imaging

TR18 (AIIM TR18-1989)
Equipment Safety

TR21 (AIIM TR21)
Optical Disk Package Labeling

TR22 (AIIM TR22)
Content of Product Specification Sheets for Optical Based Image Management Equipment and Software (*in development*)

TR25 (AIIM TR25-1990)
Guidelines Optical Imaging Systems Used for Public Records

TR26 (AIIM TR26)
Tutorial on Image Resolution (*in development*)

TR27 (AIIM TR27-1991)
Electronic Imaging RFP Guidelines

TR28 (AIIM TR28-1991)
Expungement Correction of Amendment of Information Recorded on Optical Write Once Read Many (WORM) Systems

AIIM Tools

X440 (ANSI/AIIM MS44)
Scanner Test Target Set

X441 (ANSI/AIIM MS44)
AIIM Scanner Target

Other Standards

CCITT 84a-1984
Recommendation T.4, Standardization of Group 3 Facsimile Apparatus for Document Transmission

CCITT 84b-1988
Recommendation T.6, Facsimile Coding Schemes and Control Functions for Group 4 Facsimile Apparatus

Z39.68
Related Standards for CD-ROM

ISO/CD 9171-1
Information Processing Systems—130mm Optical Disk Cartridge—Write-Once—Part 1: Unrecorded Optical Disk

ISO/CD 9171-2
Information Processing Systems—130mm Optical Disk Cartridge—Write-Once—Part 2: Recording Format

ISO 9660-1988
Volume and File Structure of CD-ROM for Information Interchange

ISO/IEC 10149
Information Processing Systems Data Interchange on Read-Only 120mm Optical Data Disks (CD-ROM)

Appendix

D

Sample Imaging RFP

WAUKESHA COUNTY
PURCHASING DEPARTMENT
615 W. MORELAND BLVD.
WAUKESHA, WI 53188

Formal Proposal No. F-9231PS
Date of Proposal 10-14-92
Buyer Susan M. Connelly, C.P.M.
Phone 548-7888
Page 1 of 9

1. REQUEST FOR PROPOSAL, INSTRUCTIONS, TERMS, CONDITIONS

This RFP is issued on behalf of Waukesha County by the Department of Purchasing which is the sole point of contact during the procurement process.

Sealed proposal for furnishing supplies, equipment, work, material, labor, services, or material and labor combined as specified below will be received in the office of the Waukesha County Purchasing Department, 615 W. Moreland Blvd., Waukesha, WI 53188 no later than 10:15 a.m. C.S.T. on

November 3, 1992

ALL PROPOSALS RECEIVED AFTER THIS DATE AND TIME WILL BE REJECTED

2. PROPOSAL SUBMISSION:

The proposal must be SEALED, addressed to the Waukesha County Purchasing Department, 615 W. Moreland Blvd., Waukesha, WI 53188. ____ copies of your proposal must be submitted. Your proposal envelope must be plainly marked in lower left hand corner of the envelope as follows:

F-9231PS	November 3, 1992	RFP-Preparation for Imaging Technologies
Proposal No.	**Opening Date**	**Item**

Any proposal received in an envelope not properly and clearly marked as specified above may result in the proposal being rejected. It is the vendor's responsibility to comply with the above in order to assure its inclusion at the proposal opening.

IMPORTANT

X Your proposal must meet the plans and/or specifications set forth.

X Proposals must be submitted on the forms provided. If forms provide insufficient space, vendors may attach a supplemental sheet(s). However such additional sheet(s) shall also be signed.

X The proposal must be signed by a person or persons authorized to legally bind your firm or it will not be considered. All requirements of item #45 (last page) must be fully executed in order for your proposal to be considered.

ANY PROPOSALS NOT CONFORMING WITH THESE REQUIREMENTS WILL BE REJECTED.

3. DESCRIPTION OF WORK:

Request for Proposal for Preparation for Imaging Technologies in accordance with Request for Proposal #F9231PS dated 10/9/92.

All proposals will be opened publicly however, only the names of the firms proposing will be read publicly at the date and time indicated above.

WCP1891

4. PLANS AND/OR SPECIFICATIONS

Request for Proposal in accordance with the attached Request for Proposal #F-9231PS dated 10/9/92.

5. SPECIAL CONDITIONS:

The following special conditions apply to this proposal and the award of the contract following the receipt of proposals.

Award will be made to the responsive and responsible proposer who complies with the provisions set forth in the Request for Proposal and who obtains the highest score in accordance with the evaluation criteria (Reference Paragraph G, Page 10.)

Any special conditions stated above shall take precedence over any conflicting provisions stated elsewhere in this document.

6. PERFORMANCE BOND REQUIRED: $____________ (Upon Award of Contract)

INSTRUCTIONS, TERMS AND CONDITIONS FOR FORMAL PROPOSAL

7.
(a) **PERFORMANCE BONDS** – If required, a performance bond or a single letter of bank credit in the amount specified in Item #6 is to be provided by the successful proposer to the Waukesha County Purchasing Department within ten (10) calendar days after award of the contract. Failure to do so can make the contract voidable at the County's discretion and where the proposal is covered by a certified check or bond such security shall become the property of Waukesha County as liquidated damages. Performance bonds must be approved by the Risk Manager prior to the commencement of any work.
(b) **INSURANCE** – If applicable, an insurance certificate, which meets the County's requirements, as specified in the proposal document shall be provided to the Purchasing Department with the proposal or as stated in the proposal/specifications. Failure to do so can make the contract voidable at the County's discretion. Said insurance certificate must be approved by the Risk Manager prior to the commencement of any work. The contractor shall not allow any subcontractor to commence work until the insurance required of the subcontractor has been obtained and approved.
(c) The performance bond and the insurance certificate must be issued by companies licensed to do business in the State of Wisconsin or signed by an agent licensed by the State of Wisconsin. Waukesha County will be named as additional insured with respect to general liability coverage. Should any insurance policy be cancelled before the expiration of this contract, the issuing company must provide 10 days notice to Waukesha County.

8. **SAFETY REQUIREMENTS** – All material, equipment and supplies provided to the County must comply fully with all safety requirements as set forth by the Wisconsin Administration Code, Rules of the Industrial Commission on Safety and all applicable OSHA Standards.

9. **LICENSES & PERMITS** – Vendors shall have any and all licenses and permits required to perform the work specified and furnish proof of such licenses, authorization and permits with their proposal if requested.

10. **DELIVERY F.O.B. DESTINATION** – All proposal prices must include delivery F.O.B. destination unless otherwise specified in the Request for Proposal.

11. **TAXES TO BE EXCLUDED FROM PRICE** – All proposals must be submitted without the inclusion of federal excise and Wisconsin sales taxes as the County is exempt therefrom.

12. **SIGNATURE REQUIREMENT**
PROPOSALS MUST BE SIGNED – This proposal must be manually signed by the vendor. (See Paragraph 45). All attachments, additional pages, addenda, or explanations supplied by the vendor with this proposal will be considered as part of the proposal.

13. **CASH DISCOUNTS** – All vendors must state their best cash discount on this proposal form for payments made thirty (30) days following acceptance. Discounts will be considered in determining the award.

14. **PACKAGING** – Materials shall be clearly labeled and packaged in accordance with Federal and Wisconsin requirements, and if specified in the terms and conditions of the Request For Proposal as so specified.

15. **SUBSTITUTIONS AND EQUIVALENTS** – Substitutions of equivalents of specified items may be permitted at the sole discretion of the Purchasing Department.

16. **RIGHT TO REJECT** – The Purchasing Department reserves the right to reject any or all proposals.

17. **RIGHT TO ACCEPT ALL OR PART OF PROPOSAL** – The Purchasing Department reserves the right to accept all or part of any proposal.

18. **OTHER CONSIDERATIONS** – Quantities involved, time of delivery, purpose for which required, competency of vendor, ability to render satisfactory service, and past performance will be considered in determining status as a responsible vendor.

19. **QUALIFYING VENDORS** – Prior to Award to vendor, Waukesha County Purchasing Department may require submission by vendor of proof of financial stability/ability and experience in performance of similar obligations.

20. **WITHDRAWAL OF PROPOSALS** – Proposals may be withdrawn only in total, and only by a written request addressed to the Purchasing Department prior to the award of the contract. The Purchasing Department has sole discretion to grant a request for a withdrawal of proposal.

21. **AMENDMENT OF PROPOSALS** –
(a) **By County: Proposals may be amended by the Purchasing Department in response to need for further clarification, specification and/or requirements changes, new opening dates, etc. Copies of the amendment will be mailed to prospective vendors and shall be signed by the vendor and returned as specified in the amendment.**

(b) **By Vendor: Proposals may only be amended after receipt by the Purchasing Department by submitting a later dated proposal that specifically states that it is amending an earlier proposal. No proposal may be amended after the opening.**

(c) **Upon Request of Vendor: If any of the terms and conditions prevent you from offering a proposal, consideration will be given, if possible, to a vendor request for a change. This request must be submitted to the Purchasing Department in writing ten (10) business days prior to the proposal opening, and if granted, it will require sending an addendum to all prospective vendors.**

22. **COUNTEROFFERS** – Counteroffers, changes to any terms, conditions, specifications or plans stated herein will be rejected.

23. **OFFER AND ACCEPTANCE** – The proper submission of this form by the vendor will be considered as the vendor's offer to enter into a contract in accordance with the provisions herein set forth. All proposals shall be binding for **sixty (60)** calendar days from the date of opening, unless otherwise specified under No. 5 (Special Conditions) of the proposal. If your proposal is accepted and a contract issued, then this proposal will be incorporated into a contract between the County and your firm, which shall contain all the terms and conditions agreed on by the parties hereto, and no other agreement, regarding the subject matter of this proposal, shall be deemed to exist or bind any of the parties hereto. Acceptance will take place only upon award by the Purchasing Department, execution of a contract by the proper County officials, and delivery of the fully-executed contract to the successful vendor. Acceptance may be revoked at any time prior to delivery of the fully-executed contract to the successful vendor.

24. **APPLICABLE LAW** – Except as provided herein, Wisconsin will be the forum for all disputes.

25. **ASSIGNMENT OR SUBCONTRACT** – This contract may not be assigned by the successful proposer without the written consent of the Purchasing Department. All subcontractors must also be approved by the Purchasing Department.

26. **CONTRACT CONTINGENT UPON FUNDING** – The failure of the Waukesha County Board to appropriate funds for the performance of the contract shall void the contract.

27. **RIGHT TO ORDER WITH 10%** – The Purchasing Department reserves the right to order within ten percent (10%) more or less of the quantities specified in the contract.

28. **PURCHASE ORDERS** – No shipments or performance shall be made under the contract until a purchase order has been received unless otherwise agreed by the Director of Purchasing in writing.

29. **NONDISCRIMINATION** – The successful vendor will not discriminate against any employee or applicant for employment because of age, race, creed, color, sex, sexual orientation, handicaps or national origin or ancestry. This provision must be included in all subcontracts.

30. **AMENDMENTS TO CONTRACT** – The contract may be modified only by a written amendment to the contract signed by both parties.

31. **INDEMNIFICATION** – The successful vendor will indemnify and hold harmless the County against all damages, losses, liabilities, judgments, costs and expenses arising out of the successful vendor's performance or failure to perform under the contract.

32. **DEFENSE OF SUITS** – Contractor will defend Waukesha County, its officers, agents and employees in each and every claim, lawsuit or other demand for damages filed in connection with or occurring during the course of this agreement/contract where such liability is founded upon or grows out of the acts or omissions of any agent or employee of the contractor.

33. **WAIVER** – One or more waivers by any party of any term of the contract will not be construed as a waiver of a subsequent breach of the same or any other term. The consent or approval given by any party with respect to any act by the other party requiring such consent or approval shall not be deemed to waive the need for further consent or approval of any subsequent similar act by such party.

34. **DELAYS IN DELIVERY** – Delays in delivery caused by government priority or requisition, riots, fires, sabotage, acts of God, or any other delays deemed by the Purchasing Department to be clearly and unequivocably beyond the contractors' control will be recognized by the County. The contractor may be relieved of the responsibility of meeting the delivery time as stipulated in the contract upon contractors' filing with the Purchasing Department just and true statements requesting an extension of delivery, signed by the contractor and giving in detail all the essential circumstances which, upon verification by Waukesha County, justify such action under the provisions of this section by the Purchasing Department. THE REQUEST FOR EXTENSION MUST BE FILED WITH THE PURCHASING DEPARTMENT NO LATER THAN SEVEN (7) CALENDAR DAYS PRIOR TO THE ACTUAL DELIVERY DATE. FAILURE TO FILE THIS REQUEST FOR DELIVERY EXTENSION SHALL RENDER THE CONTRACTOR LIABLE FOR THE DIFFERENCE BETWEEN THE "OPEN MARKET" AND THE CONTRACT PRICE, AND OTHER COSTS, AS APPLICABLE UNDER THE UNIFORM COMMERCIAL CODE.

35. **MATERIAL SAFETY DATA SHEETS** – It is a direct condition of this order that the vendor shall supply Waukesha County, when applicable, with a MATERIAL SAFETY DATA SHEET for the material purchased. In addition the vendor shall supply any material related to the safe use of this material and hazards associated with its use, including but not limited to installation procedures and personnel protective equipment requirements. All hazardous components shall be identified.

ALL DATA SHEETS MUST HAVE THE CORRESPONDING COUNTY PURCHASE ORDER NUMBER **CLEARLY PRINTED** ON THE FIRST PAGE. Sheets must be sent to the Purchasing Department prior to the shipment of the material. NO PAYMENTS WILL BE MADE UNTIL THE MATERIAL SAFETY DATA SHEETS ARE RECEIVED.

36. EXTENSIONS – Agreements shall run for the period indicated subject to two one year extensions. Option of extension, if mutually agreeable to the County and the contractor, shall be in writing and may be exercised within sixty (60) days of the contract expiration date, or as otherwise expressed in the proposal.

37. PAYMENT FOR CONTRACT PERFORMANCE – Upon the complete performance of this contract by the contractor, and after the acceptance of said performance by Waukesha County, the contractor shall be paid any balance then remaining due and payable under the terms of this contract for said work, material, services, items, supplies or equipment.

38. CONTRACT DEFAULT – If the contractor shall fail to fully and completely perform the contract within the time designated for the performance thereof, the contractor shall pay Waukesha County, as liquidated damages for such default, any amount of any deposit or bond required by the proposal. Failure to require such amounts, deposits or bonds shall not limit the remedies otherwise available to Waukesha County.

39. NON-INTEREST OF COUNTY EMPLOYEES AND OFFICIALS – No official or employee of Waukesha County shall have any financial interest, either direct or indirect, in the proposal or contract. No official or employee of Waukesha County shall exercise any undue influence in the awarding of the proposal or contract.

40. PROPOSAL RESULTS – Vendors may obtain information pertaining to results of proposals only after the notice of award has been released by visiting the County Purchasing Department, Monday through Friday between 8:30-11:30 a.m. and 1:00-4:00 p.m.

41. FAILURE TO SUBMIT A PROPOSAL – Failure to respond to a Request for Proposal (submission of proposal, or notice in writing that you are unable to propose, but wish to remain on the active vendor's mailing list qualifies) will be understood by the County to be a lack of interest and result in the removal of the vendor's name from the applicable vendor's list.

Formal Proposal No. F-9231PS
Date 10/14/92
Page 7 of 9

COST SHEET FOR FORMAL PROPOSAL
To the WAUKESHA COUNTY PURCHASING DEPARTMENT

42(P). The submission of a proposal shall be considered as a representation that the proposer has carefully investigated all conditions that affect, or may at some future date affect the performance of the Services covered by this solicitation, the entire area to be serviced as described in the technical specifications and other documents; and that the proposer is fully informed concerning conditions to be encountered, the character, quality and quantity of work to be performed and the materials to be furnished. Also, the proposer is familiar with all applicable State, Federal and Local regulations that in any way affect prosecution of the work or persons engaged or employed in the work. The contents of the proposal of the successful contractor shall become contractual obligations if procurement action ensues. Failure to accept these obligations in a contractual agreement shall result in cancellation of the award.

State total not to exceed cost (inclusive of all costs, travel, expenses etc.) to perform the services delineated in the Request for Proposal Preparation for Imaging Technologies

$ ______________________________ Total Fixed Fee

Important: This form must be signed (in addition to your proposal) Manual Signatures only - no facsimilies.

One copy of this form should be in an envelope with your cost proposal which is separate from the Technical Proposals (7 copies of your technical proposal must be submitted) and the envelopes must be identified in the lower left corner as follows:

F-9231PS Imaging Technologies RFP
Opening Date: 11/3/92 @ 10:15 A.M. CST. - Technical Proposal

and

F-9231PS - Cost Proposal - Preparation for Imaging Technologies RFP

NOTE: Proposals will be opened November 3, 1992 at 10:15 A.M. All proposals must be time stamped in by 10:15 A.M. CST on the opening date.

Envelopes which are not properly identified or are received after the above date and time will be rejected.

The names of the firms bidding will be the only information released at the proposal opening.

Proposals will **not** be available for review on the date of proposal opening.

Proposals will be made available for review after all interviews have been conducted and a recommendation for award has been made.

Formal Proposal No. F-9231PS
Date 10-14-92
Page 8 of 9

43. **CASH DISCOUNT:**_________% Cash discount for payment within thirty (30) days following receipt and acceptance of goods. (Discount for earlier payment will not be considered for purpose of award).

44. **DELIVERY:** Time is of the essence for all terms of the contract. Delivery no later than ________ calendar days from date of award of contract. as specified in the Request for Proposal dated 10-9-92.

NOTE: IF YOU ARE UNABLE TO MEET THIS DELIVERY SCHEDULE, YOUR PROPOSAL WILL BE REJECTED!

GUARANTEED PERFORMANCE: Failure by the contractor to adhere to the performance/delivery schedule as specified, or to promptly replace rejected or defective materials within a specified time frame, shall render the contractor liable for the difference between the "open market" and the contract price, and/or other costs as applicable under the Uniform Commercial Code. Such purchases will be deducted from contract quantities.

Formal Proposal No. F-9231PS
Page 9 of 9

WAUKESHA COUNTY
PURCHASING DEPARTMENT
BINDING SIGNATURES FOR FORMAL PROPOSAL

45. Upon award and acceptance of this proposal by Waukesha County the undersigned vendor, by its authorized persons signing below, agrees that all provisions set forth herein will become binding as part of a formal contract upon the fulfillment of all condtions precedent set forth herein.

CONTRACTOR **(To be signed by two persons authorized to legally bind your firm to this contract)**

This section must be fully completed and witnessed in order to process your proposal.

Bidder's Firm__________

Address__________

City & State__________ Zip______

Telephone No. ( )__________

Fax No. ( )__________

BY:__________
(Manual Signature Required)

Name__________
(Typed or Printed)

TITLE:__________

DATE:__________

BY:__________
(Manual Signature Required)

Name__________
(Typed or Printed)

TITLE:__________

DATE:__________

NO FAX BIDS WILL BE ACCEPTED. ANY FAX BIDS RECEIVED WILL BE REJECTED.

Witnesses: Two Witnesses to Bidder's Authorized Signature.

____
(Manual Signature Required) Date

____
(Name of Witness to be Typed or Printed)

____
(Manaul Signature Required) Date

____
(Name of Witness to be Typed or Printed)

WAUKESHA COUNTY

PURCHASING DEPARTMENT

615 W. MORELAND BOULEVARD

WAUKESHA, WISCONSIN 53188

PREPARATION FOR IMAGING TECHNOLOGIES

REQUEST FOR PROPOSAL

F9231PS

October 9, 1992

A. INTRODUCTION

1. PURPOSE

The purpose of this Request for Proposal (RFP) is to procure consulting services sufficient to:

Assess current industry imaging technology standards, and usage, and Waukesha County records management practices.

Define a strategic imaging technology architecture for Waukesha County.

Develop strategies for identifying and ranking opportunities for applying imaging technology in Waukesha County.

Plan the implementation of identified imaging application opportunities.

2. BACKGROUND

WAUKESHA COUNTY

Waukesha County is located in southeastern Wisconsin. It is the third largest county in the state with a population in excess of 310,000 people. Its growth, in relationship to residents and new business development, surpasses all other counties in the State. The County employes more than 1500 people and has a 1992 Operating Budget of more than $107 million, in addition to a Capital Project Budget that exceeds $16 million. The courthouse houses the majority of the County offices and is located at 515 W. Moreland Blvd. in the City of Waukesha, Wisconsin 53188. There are however, a variety of offices in other buildings on and off the County campus. Future plans, dictate that a new Administrative Office Building and a Mental Health Center will be built in addition to the newly constructed Justice and Law Enforcement Building which will be ready for occupancy in the later part of this year or early 1993. It is anticipated that County office space requirements will increase by 26% by the year 2010. These new facilities will alleviate most of the present crowded conditions.

RECORDS SYSTEMS

The majority of the records within the County are maintained in hardcopy paper form. File cabinets, open shelf filing units, mobile shelving and boxes are used to store the paper files. Waukesha County launched a new microfilm program two and a half years ago. Approximately $300,000 was invested in state of the art micrographic equipment. The end result was a totally self sufficient operation. While microfilming had been used as a means of record storage prior to this renovation, it was limited to a few choice departments.

V31844/dls

Over the last two years, microfilming has expanded to provide services for 13 departments. These services include handling the record retrievals from the microfilm for several of the users. The Micrographic Services Division produces 16 mm and 35 mm roll film and aperture cards depending on the applications.

While some of the records are unique to a specific operation, others are referenced by several departments. In the latter situation, the present operations often times only permit one individual to look at a given record at a time. Because of the file back procedures, a record can actually be out of the file for several hours during a day.

The Micrographic Services Division has recently proposed a County wide Records Management Program. According to the proposed plan, the new Records Management Division, under the auspices of the Waukesha County Register of Deeds, would be charged with the responsibility of adopting record retention schedules, chairing a record retention committee, controlling off-site records storage facilities, recommending and approving the purchase of any and all records storage, filing and retrieval equipment and/or systems, developing a vital records program, designing a disaster recovery program, controlling the micrographic operation, as well as monitoring the initiation, implementation, and operation of the imaging application projects undertaken.

B. **SERVICES REQUIRED FROM QUALIFIED CONSULTANT**

1. SCOPE

The scope of this engagement will require the consultant to:

a. Assess industry technical standards, Waukesha County technical standards, relevant legal standards, and the use of imaging technology in state and local government in order to define a strategic imaging technology architecture. They should address the issues of image capture, storage, retrieval, presentation, and distribution.

b. Assess the records management practices in at least five major County departments (Register of Deeds, Sheriff, Clerk of Courts, Finance, and Environmental Resources) with respects to the potential for service improvement and cost reduction through the use of imaging technology.

c. Develop business and technical strategies which define a framework within which options modernizing Waukesha County's records management practices can be considered. The strategic framework developed should:

Rank imaging technology applications according to their potential to increase the quality and reduce the cost of the records management practices currently used in the five aforementioned departments.

Specify the economic, technical, and operational feasibility of achieving each imaging application opportunity.

Be consistent with the imaging technology architecture recommended.

d. Develop a plan for implementing imaging technology which must include, but not limited to:

Identification of pilot projects which will prove the projected benefits of the application of imaging technologies.

Specification of the criteria for judging the relative success of pilot projects.

Identification of the resources (people, technology, space, etc.), and time-frames required to achieve the pilot projects identified.

Identification of subsequent imaging projects to be implemented after successful completion of the pilot imaging projects.

The design and implementation of imaging systems will be a joint venture shared by the Register of Deeds Department and the Information Systems Department, and representatives from these departments will form an Imaging Steering Committee for the initial purpose of managing these efforts and this engagement.

2. DESIRED RESULTS

a. <u>Assessment Phase</u>. Includes, but not limited to:

Identify the imaging industry standards, both current and future directions, and the successful applications of these technologies in the County government sector; specify the future of the imaging industry as a whole.

Specify the meaning and contents of imaging technologies and applications and identify their strategic and tactical benefits, particularly to County government.

Survey the five department heads and/or assigned coordinators to learn their present services and determine their future needs.

Specify their inter- and intra- departmental records process flows and how one department's records may affect the functions of another department in relationship to where the records are stored and how they may be retrieved, as well as, how they are used in daily operations.

Specify general potential applications of these technologies throughout the County (besides these five departments), particularly where other County governments have achieved cost-effective results.

b. Strategy Development Phase. Includes, but not limited to:

Specify one or more centralized and/or decentralized options for the County to best meet its imaging services.

Identify imaging hardware and software options for the County as they relate to our County's records management and computer operations and standards, as well as for our statutory requirements.

Identify potential pilot applications, and departments involved, to prove the concepts and benefits of imaging technologies.

Identify the results and (cost) benefits expected for all these applications in implementing these technologies, as well as the people, time, and technology investments expected for their implementation.

Identify any and all current and potential legal implications of using optical imaging technologies, particularly optical disk systems as it relates to governmental regulations and usages.

Specify imaging standards for the County in implementing these applications.

c. Implementation Plan Phase. Includes, but not limited to:

Specify time lines for the County to proceed in these efforts, to include pilot project(s) and long term projects.

Specify the necessary investments during each year and for each project, to include (but not limited to) people, training, space, hardware, software, development assistance.

3. DELIVERABLES.

a. A draft ASSESSMENT report to be reviewed by the Steering Committee for submission of comments and changes desired within two weeks upon receipt. A final draft will then be prepared and submitted to the Steering Committee.

b. A draft STRATEGY DEVELOPMENT report to be reviewed and finalized in the same manner.

c. A draft IMPLEMENTATION PLAN report to be reviewed and finalized in the same manner.

d. A final IMAGING TECHNOLOGIES PLAN report consisting of the contents of the finalized three previous draft reports.

e. No more than three formal presentations of the IMAGING TECHNOLOGIES PLAN, most probably to the County Executive, County Board members, and perhaps to County departmental members effected jointly.

C. <u>DURATION OF THIS ENGAGEMENT</u>

The duration of this engagement is expected not to exceed 120 calendar days from the day the contract is awarded.

D. <u>QUALIFYING CRITERIA</u>

In order to be considered for inclusion on the accepted consultant list to be developed by the Request For Proposal, the consultant must have the following minimum qualifications:

1. Must be well versed in records management practices, including but not necessarily limited to, hardcopy paper filing systems, automated retrieval systems and micrographic operations and must be able to exhibit accomplishments over at least the last 5 years.

2. It is highly desirable that the consultant is able to exhibit knowledge of county government ordinances and operations, including but not necessarily limited to, records practices and Wisconsin State Statutes governing the same.

3. Must have a minimum of 3 years practical experience in optical disk imaging technology. Has to exhibit technical and operational knowledge and be able to show successful applications in an active working environment where the consultant's input was instrumental in the initiation and implementation of the systems. This knowledge must include conversions of microfilm and aperture cards to optical disks.

4. Must be knowledgeable of industry standards established by the American National Standards Institute (ANSI), the Association for Information and Image Management (AIIM) and the Association of Records Managers and Administrators (ARMA) as they relate to imaging technology.

5. Must be able to exhibit knowledge of stand alone imaging systems, as well as, networking and tying into existing main frame computer systems.

E. PROPOSAL RESPONSES

1. **FORMAT**

The County reserves the right to reject any proposal not formatted as requested.

The vendor's response to this Request For Proposal must be formatted in separate sections as follows:

Section 1

Verify your understanding and the acceptance of the DESIRED RESULTS and DELIVERABLES sections addressing each item in the same order as presented in the RFP (sections B2 and B3.).

Section 2

Address the minimum QUALIFYING CRITERIA (section D) listing the responses in the same order as presented in the RFP (section D). Specific attention must be made on the expertise and experience in accomplishing similar engagements.

Section 3

Submit a work plan describing in detail how the tasks will be accomplished and providing a time line relating to each phase and submission of the final report.

Section 4

Furnish at least three (3) references of previous engagements on projects of similar natures. Provide a contact name and phone number. These references should only be for completed projects. Additionally, provide the names of the individuals from your firm who managed and worked on the respective completed referenced projects.

Section 5

Furnish information on your company, including such items as:

1. Lines of business

2. Management and technical expertise

3. Years in business

4. Financial information for the last 3 years consisting of, but not limited to, financial statements, insurance coverage, a Dunn & Bradstreet Report or similar report on the company's financial condition

5. Statement on business stationery signed by the owner or authorized official indicating that their firm is not currently and has not undergone bankruptcy proceedings

Section 6

Furnish resumes of senior technical members of your company who would be expected to be available for use on the various phases of the projects. Identify the total number of hours or percent of time each will devote to the proposed project at Waukesha County including their hourly rate and what each individual's role will be in the completion of the project.

Section 7

Identify any subconsultants or associates that would be involved in this project and define their roles.

Section 8

Furnish consulting fee structure and submit a "total project not to exceed cost" which must include all costs (travel, expenses, materials etc.).

Section 9

Furnish any other information which you consider relevant to the evaluation of your response to this RFP.

NOTE: VENDORS SHOULD BE AWARE OF THE FOLLOWING:

The vendor proposal is a public document and must be made available to the public and press. Confidential and proprietary material must be avoided or a special request must be made to the Waukesha County Purchasing Department attention: Susan Connelly) prior to submitting them. The County reserves the right to refuse acceptance of proprietary or confidential information which it considers irrelevant or superfluous. If parts of the response are accepted as proprietary or confidential, the response must be accompanied by the forms given in appendix A.

2. **SUBMISSION**

Seven (7) copies of the technical response to this Request For Proposal must be submitted in a sealed envelope no later than **10:15 PM CST on November 3, 1992** to the following address:

Waukesha County Purchasing Department
615 W. Moreland Blvd.
Waukesha, WI 53188

The envelopes must be identified in the lower left hand corner as follows:

F9231 PS - Imaging Technologies RFP

NOTE: Proposers <u>must</u> submit their fee in a separate envelope unattached to their proposal and identify it as **Cost Proposal - F9230 PS - Imaging Technologies RFP**

Envelopes which are not properly identified or are received after the above date and time will be rejected.

3. **CLARIFICATIONS**

Questions concerning any of the items addressed in this Request For Proposal will be answered prior to the date for submission of responses. Questions must be made in writing to Susan Connelly, Director of Purchasing at the previously stated address and must arrive on or before October 21, 1992. The questions and appropriate answers will be furnished to all vendors who have received this RFP. The County reserves the right to reject answering any question received late and not submitted in writing.

4. **RESTRICTIONS**

Only vendors who are responding as principals acting as agents on their own behalf will be considered for evaluation.

F. **<u>SELECTION PROCESS</u>**

Responses to this Request For Proposal will be evaluated on by using the following criteria:

1. Experience and qualifications of the proposer and staff available for assignment to the upcoming project.
2. Exhibited organizational ability to successfully complete projects as assigned on a timely basis.
3. Demonstrated history of successes in completing projects of similar nature and scope.
4. Availability to start project
5. Fee structure

NOTE: The County reserves the right to request an interview of any or all consultants submitting proposals to clarify submissions and/or seek additional information. Interviews will be at the proposer's expense. The award may be made without discussion of offers of any proposals received. Therefore, offerers are cautioned that proposals should be sumitted initially on the most favorable terms, from both a technical and cost standpoint.

G. **EVALUATION OF PROPOSALS**

Award will be made to the responsible proposer whose proposal is determined to be the most advantageous to Waukesha County based upon the following criteria:

1. The completeness of the proposal, including scope, approach and detailed work plan.

   0-25 points

2. Firm experience in related consulting services.

   0-20 points

3. Individual experience and education relevant to services required for all personnel that would be assigned to the project.

   0-25 points

4. The degree to which the proposal satisfies all requirements of the request for proposal.

   0-5 points

5. Not to exceed predetermined cost limitations.

   0-25 points

Maximum total points 100 points

NOTE: No official or employee of Waukesha County shall have any financial interest, either direct or indirect, in the proposal or contract. No official or employee of Waukesha County shall exercise any undue influence in the awarding of the proposal or contract.

H. **NOTIFICATION**

It is anticipated that the consulting firm selected for this project will be notified in writing within 4 weeks after the opening date.

The names of the firms proposing will be the only information released at the proposal opening.

I. **MISCELLANEOUS**

This request does not commit Waukesha County to award any contract nor to pay any costs incurred in the submission of a proposal.

Questions:
The County reserves the right, in its sole discretion, to reject any or all responses, issue addenda, request clarification, alter the nature and/or scope of the proposed study, request resubmission and/or discontinue this process. Waukesha County also reserves the right to make no award.

V31844/dls

All proposers should use this written document and its attachments as the sole basis for proposal at this time. The county will not be responsible for oral interpretation given by any County employee, representative or others. The sole contact for questions regarding this Request for Proposal is Susan Connelly, Director of Purchasing. (See #3 Clarification on Page 9.)

Period before Withdrawal: All proposals shall be firm offers and may not be withdrawn for a period of ninety (90)days following the proposal submission due date.

APPENDIX A

DESIGNATION OF CONFIDENTIAL AND PROPRIETARY INFORMATION SERVICES

Material submitted in response to the Request for Proposal No. ________________ for __________________________ Waukesha County may be considered confidential or proprietary and therefore, under the Wisconsin Open Records Law may be kept confidential. As such, the proposer asks that certain pages, as indicated below, of this proposal be treated as confidential material and not released without the proposer's written approval.

CHECK ONE:

( ) I am providing the following information with the express understanding that it is being submitted to the County under a pledge of confidentiality and can only be released to third parties with my written permission. I would not have submitted this information had Waukesha County not pledged to keep it confidential.*

or

( ) I have no objection to the release to third parties of the following information.

The proposer requests that the following pages not be released:

Section	Page #	Topic

*NOTE: Proposers are cautioned that the ENTIRE PROPOSAL MAY NOT FALL WITHIN THE CONFINES OF THE PLEDGE OF CONFIDENTIALITY. Only that information that would not have been submitted save for pledge of confidentiality falls within the exception of public inspection.

IN THE EVENT THE DESIGNATION OF CONFIDENTIALITY OF THIS INFORMATION IS CHALLENGED, THE UNDERSIGNED HEREBY AGREES TO PROVIDE LEGAL COUNSEL OR OTHER NECESSARY ASSISTANCE TO DEFEND THE DESIGNATION OF CONFIDENTIALITY.

Failure to include this designation in the proposal response may mean that all information provided as part of the bid/proposal response will be open to examination and copying. Waukesha County considers other markings of confidential in the proposal document to be insufficient. The undersigned agrees to hold Waukesha County harmless for any damages arising out of the release of any materials unless they are specifically identified above.

________________________________ ______________________

Signature (Authorized Representative) Telephone Number

Page 12 of 13

V31844/dls

Bibliography

1991-1992 Information Management Sourcebook. 1991. Silver Spring, MD: Association for Information and Image Management.

1992 Electronic Imaging Directory. 1992. Potomac, MD: Phillips Publishing Inc.

Abi, Ray. 1990. *Image Processing Requirements Analysis.* Stamford, CT: Unitech International Corporation, 1990.

Abi, Ray. 1990. *Image Processing: An Integrated Approach to Information Processing.* Stamford, CT: Unitech International Corporation.

Abi, Ray. 1990. *Justifying Image Processing Systems.* Stamford, CT: Unitech International Corporation, 1990.

AIIM White Paper. 1992. Silver Spring, MD: The Information And Image Management Industry Association for Information and Image Management.

Avedon, Don. 1991. *Electronic Imaging: What It's All About*, slide presentation. Potomac, MD: Avedon Associates, Inc., 1991.

Avedon, Don. 1992. *Introduction to Electronic Imaging.* Silver Spring, MD: Association for Information and Image Management.

Badler, Mitchell, ed. 1991. *1991-1992 International Imaging Source Book*. Larchmont, NY: Microfilm Publishing, Inc.

Berg, Brian and Judith Paris Roth. 1989. *Software for Optical Storage*. Westport, CT: Meckler.

Cinnamon, Barry and Richard Nees. 1991. *The Optical Disk: Gateway to 2000*. Silver Spring, MD: Association for Information and Image Management.

Cinnamon, Barry. 1988. *Optical Disk Document Storage & Retrieval Systems.* Silver Spring, MD: Association for Information and Image Management.

D'Alleyrand, Marc. 1989. *Image Storage and Retrieval Systems*. New York, NY; McGraw-Hill.

Deloitte and Touche. 1990. *Comparison of Commercially Available Optical Based Document Imaging.* Los Angeles, CA: Deloitte and Touche.

Deloitte and Touche. 1991. *Information and Image Management: The State of the Industry 1991.* Silver Spring, MD: Association for Information and Image Management.

Eaton, Nancy, Linda Brew McDonald, and Mara Saule. 1989. *CD-ROM and Other Optical Information Systems.* Phoenix, AZ: Oryx Press.

Electronic Document Imaging Systems: User Implementation Analysis. 1991. Delran, NJ: Datapro Research Corporation.

Fruscione, James. 1988. *Conceptual Design Guideline for Optical Disk Document Management Systems.* Silver Spring, MD: Association for Information and Image Management.

Fruscione, James. 1989. *Feasibility Studies for Document Management Systems.* Silver Spring, MD: Association for Information and Image Management.

Fruscione, James. 1990. *Planning for Image Acquisition, Management, Support, and Control*. Silver Spring, MD: Association for Information and Image Management.

Hendley, Tony, John Prichard, and Simon Perkins. 1990. *Document Image Processing Reference Set, vols. I-III*. Manchester, England: National Computing Centre Ltd.

Jordahl, Greg. 1990. *Plugging into the FAX Track*. Silver Spring, MD: Association for Information and Image Management.

Kalthoff, Robert. 1990. *Buying Electronic Image Management Systems*. Cincinnati, OH: Strategy Inc.

McCormick, John. 1990. *A Guide to Optical Storage Technology.* Homewood, IL: Richard Irwin Inc.

Porter-Roth, Bud. 1989. *Proposal Development: How to Respond and Win the Bid.* Grants Pass, Oregon: Oasis Press.

Ranade, Sanjay and James Ng. 1989. *Systems Integration for Write-Once Optical Storage.* Westport, CT: Meckler.

Ranade, Sanjay. 1990. *Optical Storage Jukebox Technology: A Systems Integrator's Perspective.* Westport, CT: Meckler.

Roth, Judith Paris. 1990. *Case Studies of Optical Storage Applications.* Westport, CT: Meckler.

Roth, Judith Paris. 1990. *Converting Information for WORM Optical Storage*. Westport, CT: Meckler.

Roth, Judith Paris. 1991. *Rewritable Optical Storage Technology*. Westport, CT: Meckler.

Saffady, William. 1989. *Personal Computer Systems for Automated Document Storage and Retrieval.* Silver Spring, MD: Association for Information and Image Management.

Saffady, William. 1990. *Optical Disk vs. Magnetic Storage*. Westport, CT: Meckler.

Saffady, William. 1990. *Optical Storage Technology 1990-91: A State of the Art Review.* Westport, CT: Meckler.

Sant, Thomas. 1992. *Persuasive Business Proposals*. New York, NY: American Management Association.

Schantz, Herbert. 1991. *Optical Digital Imaging Text Systems*. Silver Spring, MD: Association for Information and Image Management.

Thiel, Thomas, Scott Gielda, and Ann Lancaster. 1992. *CD-ROM Mastering For Information and Image Management.* Silver Spring, MD: Association for Information and Image Management.

Wagemann, Peter C. 1989. *Handbook of Optical Memory Systems.* Newton, MA: Optical Disk Institute.

Wetzler, Fred. 1989. *Desktop Image Scanners and Scanning.* Silver Spring, MD: Association for Information and Image Management.

Williams, Robert, ed. 1987 with updates. *Legality of Optical Storage*. Chicago, IL: Cohasset Associates Inc.

Glossary

Definitions which have been approved by the International Standards Organization are designated *(ISO)*. Certain terms, the use of which are protested, are noted as deprecated. Trademarks and trade names are not included.

A/D Abbreviation for analog to digital; usually found in the form *A/D converter*, also abbreviated *ADC*.

access time (1) The time it takes to get an instruction or a unit of data from computer memory to the processing unit of a computer. (2) The time it takes to get a unit of data from a direct-access storage device to computer memory.

acuity Keenness or sharpness of perception.

address (1) The coded representation of the destination of data or of their originating terminals. Multiple terminals on one communication line, for example, must have unique addresses. Messages carry an address before their text to indicate the destination of the message. (2) Either a specific location in computer memory, a specific location on a direct-access storage device, or a relative location.

addressability The capability of placing intelligence at a specific point within an area. See also *addressable capacity.*

addressable capacity The number of addressable positions within a specified image or display space as follows: addressable horizontal positions by addressable vertical positions.

addressable horizontal positions The number of positions within an image at which a full-length vertical line can be placed.

addressable point Any point within an image that can be addressed.

addressable vertical positions The number of positions within an image at which a full-length horizontal line can be placed.

AIIM See *Association for Information and Image Management.*

algorithm A mathematical formula or procedure.

American National Standards Institute (ANSI) A nongovernmental organization chartered to oversee U.S. standards activities and to coordinate the U.S. role for international standards activities.

American Standard Code for Information Interchange (ASCII) The ASCII binary-coding scheme consisting of 128 eight-bit patterns (seven bits plus a parity check bit) for printable characters and control of equipment functions.

analog transmission The transmission of electronic signals analogous to tonal variations constituting the content of a document page or any form of original graphics; the representation of visual tonal variations at the input of a scanning system by proportional variations in strength or frequency of a transmitted electrical current.

analog A continuously variable measurement of physical phenomenon. These representations bear some physical relationship to the original quality; usually electrical voltage, frequency, resistance, or mechanical translation or rotation.

ANSI See *American National Standards Institute*.

aperture card scanner A device for scanning microfilm in aperture cards. See also *aperture card*, *reader-scanner* and *scanning*.

aperture card A card with one or more apertures specifically designed for mounting or inserting microfilm before or after imaging. (ISO)

applications software Programs designed to perform a user-specific task. Examples of applications software include word processing and full-text search and retrieval.

archival media A recording material that can retain information indefinitely so that it can be retrieved without significant loss when properly stored. This is not a term to be used in material or system specifications. See also *LE designation* and *life expectancy*.

artificial intelligence (AI) Computer programs that perform functions, often by imitation, that are normally associated with human reasoning and learning.

ASCII See *American Standard Code for Information Interchange*.

aspect ratio The ratio of width to height of a whole image, document, reproduction format, etc.

Association for Information and Image Management (AIIM) An association in the micrographic and electronic information and imaging field that includes company and individual memberships.

asynchronous A mode of data transmission in which each character is transmitted as a separate message often identified by start and stop bits. Contrast with *synchronous*.

attribute data Textual data describing qualitative or quantitative characteristics of, for example, a group of documents or geographic features. Attribute data relate to themes or types of information.

automatic answering The capability of a terminal, modem, computer, or similar device to respond to an incoming call on a telephone line and to establish a data connection with a remote device without operator intervention.

automatic document feeder A mechanical device used to feed documents sequentially into a camera or scanner. It may also position the document. (ISO)

automatic send A communications capability whereby the communicating system can send out a message automatically in an unattended mode.

automatic send/receive Capabilities that permit a message to be held offline for later transmission.

automatic speed selection In facsimile, a feature that verifies that the receiving unit is operating at the same speed as the sending unit. If not, the sending or receiving unit switches to the proper speed.

backbone network A transmission facility or arrangement of facilities designed to interconnect lower-speed distribution channels or clusters of dispersed users or devices.

backup A means of protecting valuable information and data. Backup may take the form of duplicating tapes or disks on which information is stored providing a system with an alternate power source to protect data in volatile memory in the event of a power failure or providing a redundant system.

bandwidth (1) The number of hertz expressing the difference between the lower and upper limiting frequencies of a frequency band. (2) The width of a band of frequencies. (3) The maximum number of information units (bits or characters) capable of traversing a communications path per second. See also *channel.*

barcode scanner A device used to read such barcodes as the Universal Product Code by means of reflected light.

barcode An array of vertical rectangular marks and spaces in a predetermined pattern.

barcode symbol A machine-generated and readable representation of data (usually numeric) in the form of a printed series of contrasting parallel bars of various widths, spacings, and heights.

baud A unit of transmission speed equal to the number of signal events per second. In asynchronous transmission, the unit of signaling speed corresponding to one unit interval per second; that is, if the duration of the unit interval is 20 milliseconds, the signaling speed is 50 baud. Technically, baud is the same as bits per second when, and only when, each signal event represents exactly 1 bit (which is rarely true), but in casual, nontechnical usage, baud is often used to mean bits per second.

BER See *bit error rate.*

binary code Code which represents information as a sequential series of zeros and ones.

binary digit (bit) A bit represents the binary code (0 or 1) with which the computer works. A bit can take the form of a magnetized spot, an electronic impulse, a positively charged magnetic core, etc. A number of bits together are used to represent a character in the computer. See also *binary code* and *binary notation.*

binary large objects (BLOB) Recording, storing, retrieving and transmitting multimedia (images, video, voice, graphics, and data) that are made up of large numbers of bytes.

binary notation Any notation that uses two different characters, usually the binary digits zero and one. (ISO)

binary Pertaining to a system of numbers with a base of two.

bisynchronous A set of operating procedures for the synchronous transmission of binary coded data.

bit string Group of arbitrarily arranged binary bits.

bit See *binary digit.*

bit-error-rate (BER) A measure of the frequency of erroneous bits expressed as the number of bits encountered before one erroneous bit is found.

bitstream Refers to a bit sequence without regard to groupings by character.

bitmap A method of representing images by assigning an individual memory location for each picture element (pixel). See also *digital image* and *pixel.*

bitmapped display A display screen on which a character or image is generated or refreshed from a bitmap.

bitmapped image A representation of image data where each pixel has a corresponding memory element. See also *digital image*, *image*, *memory*, and *pixel.*

bitonal Having picture elements (pixels) that are only one bit each. A bitonal image has two intensity values (zero and one), corresponding to black and white.

bits per inch (bpi) A measure of the density of information stored on media.

bits per pixel A measure of the number of bits contained in a given pixel. See also *bitonal.*

bits per second (bps) A measure of the speed of transmission in information systems. See also *baud.*

black and white scanner A scanner that interprets scanned data as black or white, but with additional software, can perform electronic screening, dotting, or dithering to produce simulated grayscale pixel configurations. See also *dithering*, *grayscale*, and *pixel.*

BLOB See *binary large objects.*

block error correction During CD-ROM premastering, the application of data recovery methods to a physical block of data to ensure the recovery of all data.

block A set such as documents, words, characters, or digits handled as a unit.

Boolean search A search strategy for selected information that uses AND, OR, and NOT functions.

bps See *bits per second.*

broadband absorption Absorption of a wide spectrum of light wavelengths.

broadband A general term referring to a wide range of frequencies, a device that can accept a wide range of frequencies, etc. See also *bandwidth.*

browsing A system's ability to find an undefined feature or set of features in a database or document in a collection.

bubble In optical memory, those vesicles which are formed by a laser in an optical recording medium. Where vesicles exist, light is reflected away from the reading optics, thereby creating a contrast to the vesicle-free areas of the medium surface.

bug An error in a computer program.

bus network A local area network in which there is only one path between any two data stations, and in which data transmitted by any station is concurrently available to all other stations on the same transmission medium. See also *network*, *network topology*, and *node*.

bus (1) In computer architecture, an internal path by which information travels to and from various components of a system. (2) In data communications, a network topology in which stations are arranged along a linear medium (e.g., a length of cable). See also *network topology*.

business process reengineering (BPR) Also known as business process redesign. See *reengineering*.

byte (1) A group of bits, processed or operating together. (2) A term used to describe one character of information. The most common byte is eight bits long. A byte allows 256 different possible combinations of eight binary digits. See also *gigabyte*, *kilobyte*, *megabyte*, *petabyte*, and *terabyte*.

cache storage Temporary storage, usually magnetic, that provides rapid access to data or instructions.

CALS See *computer-aided acquisition and logistics support*.

cartridge (1) In electronic imaging, a case containing an optical disk. See also *case*. (2) In micrographics, a single-core container, used to enclose processed roll microfilm, which must be completely rewound to allow removal from the apparatus. (ISO)

CAV See *constant angular velocity*.

case The housing for an optical disk that protects the disk and facilitates disk interchange.

cathode ray tube A signal-to-image converter tube in which a well-defined and controllable beam of electrons is produced and directed onto a surface to give a visible or otherwise detectable display or effect. (ISO)

CCD See *charge-coupled device*.

CCITT See *International Telegraph and Telephone Consultative Committee*.

CD-ROM See *compact disk-read-only memory*.

CD-ROM mastering process The creation of the first recording (the master) in the CD-ROM replication process.

CD-ROM premastering See *premastering*.

CD-WORM See *compact disk-write once read many*.

central processing unit (CPU) The main component of a computer that includes the circuits controlling the interpretation and execution of instructions.

channel The path or circuit along which information flows.

character One of a set of symbols that may be arranged in ordered groups to express information.

character recognition The identification of graphic characters by automatic means. See also *magnetic ink character recognition*, *optical character recognition*, and *optical scanning*.

characters per second (cps) A unit of measure equal to the number of characters a device is capable of displaying or printing in one second.

character string Any group of characters acted upon in a computer system as though it were a single unit.

character transfer rate The rate at which characters are transferred from one place to another, for example, magnetic tape to computer, computer to magnetic tape, or computer to microfilm.

charge-coupled device (CCD) A semiconductor that can collect, store, and move charges in packets. An imaging charge-coupled device responds to light and provides electrical signals.

charge-injection device (CID) A solid-state electronic light detection mechanism that can be used in a scanner.

check bit See *parity bit*.

CID See *charge-injection device*.

circuit A means of two-way communications between two points.

CLV See *constant linear velocity*.

code (1) The unique bit configuration describing a symbol or character. (2) A system of symbols representing rules for handling the flow or processing of information.

code density The number of code elements that can appear per unit of length.

COLD See *computer output to laser disk*.

color scanning The process of detecting and separating colors and assigning a digital code to the hue and intensity of each pixel. See also *digital* and *pixel*.

communications network The connection of geographically separated communications devices via transmission channels.

communications protocol The exchange of a special sequence of control characters between a computer and a remote terminal in order to manage communications.

compact disk A read-only optical disk available in formats for audio, data, and other information.

compact disk-read-only memory (CD-ROM) An optical disk that is created by a mastering process and used for reading. See also *master*.

compact disk-write once read many (CD-WORM) An optical disk that is user-written and then available for reading. (ISO)

compatibility A characteristic of data processing equipment by which one machine may accept and process data prepared by another machine without conversion or code modification.

compound document A collection of independent data types such as text, graphics, scanned images, and sometimes voice.

compression See *data compression.*

compression ratio The relationship of the total bits used to represent the original to the total number of encoded bits. See also *data compression* and *data decompression.*

computer-aided acquisition and logistic support (CALS) A U.S. Department of Defense initiative supporting the electronic interchange of data and documents between contractors, government agencies, and end-users.

computer output to laser disk (COLD) A technique for the transfer of computer-generated output to optical disk, such that it can be viewed or printed without use of the original program.

constant angular velocity (CAV) A technique enabling data recorded with a variable linear density to be read, whereby the speed of rotation of the disk remains constant.

constant linear density A characteristic of an optical disk on which similar bits are of the same length regardless of their position in relation to the center of the disk.

constant linear velocity (CLV) A technique enabling data recorded with a constant linear density to be read, whereby the rotation speed of the disk varies in inverse ratio to the radial position of the reading beam.

constant voltage transformer (CVT) An electrical device that maintains constant voltage output, although input voltage may vary.

Consultative Committee for International Telegraph and Telephone (CCITT) See *International Telegraph and Telephone Consultative Committee.*

control character A character whose occurrence in a particular context initiates, modifies, or stops a control function. (1) A control character may be recorded for use in a subsequent action. (2) A control character is not a graphic character but may have a graphic representation in some circumstances. (ISO)

controller The central control processor in a computer system, such as a document filing system, a CAD/CAM system, or a word processing system.

conversational mode The mode of a data-processing system in which the user of a terminal carries on a dialogue with the computer such that each unit of input entered by the user elicits a prompt response from the computer. See also *terminal.*

conversion The process of changing a physical document to another form, such as paper to microform or paper to bitmap.

coordinate data Data expressed in coordinates that specify an addressable point on a surface or grid (such as a CRT display or microfiche).

cps See *characters per second.*

CPU See *central processing unit.*

crater A mark made by transverse displacement of the material of an optical recording medium. A crater is distinct from a hole in that a crater has a raised "lip" around its perimeter.

CRT See *cathode ray tube.*

CRT flying spot scanner A scanning mechanism that uses a CRT-generated beam of light to raster-scan a stationary document mounted on a flatbed platen. Through lens optics, the reflected beam of light is focused onto a photomultiplier and converted into an electrical signal. High resolution is a primary characteristic of this method. See also *cathode-ray tube.*

CVT See *constant voltage transformer.*

cycle A time interval in which a characteristic or repeated event occurs. In a graphics system, most processing is said to run in machine cycles.

cyclic redundancy code (CRC) A simple form of algebraic error-correcting code.

D/A See *digital to analog converter.*

data bank A comprehensive collection of data.

data bus Computer hardware that transmits coded data in either parallel or serial mode between processors or other hardware components in a computer system.

data compression The conversion of a digital image to a lower number of bits for storage. For example, a series of zeros or ones could be counted and replaced with a code that represents the number of zeros or ones in that position. See also *algorithm, data decompression, Group 3, Group 4*, and *Huffman coding.*

data decompression The regeneration of a bitmap from a compressed representation. See also *data compression, Group 3, Group 4*, and *Huffman coding.*

data element The smallest unit of data that has meaning in describing information. Synonymous with *data item* or *field.*

data file A set of related data in which all records are organized alike and can be accessed under a single name.

data input unit A component of a COM recorder that receives data from the computer (online) or from peripheral equipment (offline). (ISO)

data item See *data element.*

data The representation of facts, concepts, or instructions in a formalized manner suitable for communication, interpretation, or processing by humans or computers.

database management system (DBMS) A set of programs designed to organize, store, and retrieve machine-readable information from a computer-maintained database or data bank.

database (1) A collection of digitally stored data records. (2) A collection of data elements within records, within files, that have relationships with other records within other files.

decentralized (computer) network A collection of systems with network control functions distributed over several network nodes.

decode To interpret information that has been encrypted.

decompression See *data decompression.*

dedicated line A communications link permanently assigned between two points, not part of a switched network; also called a private line.

defect density The fraction of the total active area of a medium obscured by defects.

defect distribution The geographic distribution of errors on an optical disk measured along its tracks and radially, across its tracks, in terms of the occurrence of successive *n*-bit errors. Defect distribution concerns the media in both an unrecorded and recorded state; it does not address focus or track servo defects.

defect management In real time, defect management refers to automated programs for altering read or write power, focus, or tracking when difficulties are detected and for the decision to abandon sectors of high error content. In batch mode, defect management refers to guidelines for disk rewrite or retirement.

defect Some irregularity in a medium that disturbs its ability to carry recorded data. Defects in an optical disk include bumps, depressions, dirt, missing, and damaged heat-sensitive material. Because of the high density of optical recording, even micrometer-sized defects are important.

demodulation The process that extracts information (digital or analog) from a carrier signal so that the transmitted information may be used.

dense (1) The relation of amount of text to non-text areas on a document. (2) The number of bits in a given area of the recording medium.

device drivers Software that tells the computer how to communicate to the optical disk or other drives. These software programs are needed for any microcomputer to communicate with external devices, such as a CD-ROM drive.

digital data Data represented by binary codes.

digital image Image composed of discrete pixels of digitally quantized brightness or color. See also *pixel*.

digital scanner An optical reader that scans and converts images into digital form.

digital to analog converter (D/A) A generic term for any device that changes digital (binary) pluses into continuous wave (analog) signals. Modems are D/A converters.

digital transmission The transmission of information in the form of electrical representations of the binary digits (bits) zero and one.

digital The use of binary code to record information. Information can be text in a binary code such as ASCII, images in bitmapped form, or sound in a sampled digital form or video. Recording information digitally has many advantages over its analog counterpart, mainly ease in manipulation and accuracy in transmission.

digital/analog (D/A) See *digital to analog converter*.

digitization The use of a scanner to convert documents (on paper or microforms) to digitally coded electronic images suitable for magnetic or optical storage.

digitize The process of representing images as a matrix of picture elements (pixels), where each pixel is small enough that it can be considered homogeneous and represented by a numeric value, often just zero or one. See also *binary* and *binary code*.

digitizer A device to convert analog signals or documents into digital code, such as a document scanner.

digitizing The conversion of an image or signal into binary code. See also *analog* and *digital*.

diode An electronic device that permits electrical current to flow in one direction in order to inhibit current flow in the opposite direction.

direct read after write (DRAW) The process of reading and checking recorded information automatically after writing (recording), normally in the next revolution of the disk.

direct read during write (DRDW) The process of reading and checking recorded information automatically as it is written (recorded), in the same revolution of the disk.

disk drive See *drive.*

disk size The nominal outside diameter of the disk without cartridge, including a rim or ring (if any).

disk, grooved Disks that come from the factory with either surface relief features within the optical depth of focus of the recording layer or reflectivity changes placed within the recording layer along the tracks, to provide positional feedback to the radial tracking servo.

disk, hub The centering feature on the disk which interacts with the spindle to provide radial and vertical centering.

disk A round, flat recording medium that consists of a substrate(s) with one or more layers deposited on the surface(s) onto which information can be recorded and played back when the disk is loaded in a disk drive.

display A visual presentation. See also *cathode-ray tube.*

dithering A scanning technique to simulate digital gray areas by intermixing black and white pixels. Dithering creates the illusion of a continuous-tone image. See also *digitizing*, *image,* and *pixel.*

doc prep See *document preparation.*

document preparation The steps necessary to ready documents for filming or scanning, such as removing paper clips, staples, and bindings, and sorting by categories.

document retrieval system A system that searches, finds, and presents to the user a specified document or a complete copy of the document instead of just a citation or reference.

document sampling A random sample of documents from a collection of documents.

document (1) A medium and the data recorded on it for human use, such as a report sheet or a book. (2) By extension, any record that has permanence and can be read by a person or machine.

dot generator See *dot matrix.*

dot matrix An array of points of ink, light, or similar image-forming elements that are used to form alphanumeric characters.

dot pitch The spacing between picture elements (pixels) in a display; measured in millimeters.

dots per inch (dpi) A measure of output device resolution and quality, e.g., the number of pixels per inch on display device. Dpi measures the number of dots horizontally and vertically.

dots per millimeter (dpm) A measure of resolution, e.g., number of pixels per millimeter on a display device.

downtime Any period of nonoperation of equipment when the equipment is supposed to be operating.

dpi See *dots per inch.*

dpm See *dots per millimeter.*

DRAW See *direct read after write.*

DRDW See *direct read during write.*

drive A machine for reading and, when possible, writing a data storage medium (disk, tape, card, or otherwise). A disk can be optical, magnetic, or some other medium.

e-mail See *electronic mail.*

EBCDIC See *extended binary-coded decimal interchange code.*

ECC See *error correction code.*

EDAC See *error detection and correction.*

EDI See *electronic data interchange.*

EIM See *electronic image management.*

electronic data interchange (EDI) A method of exchanging information as data formatted for computer processing, rather than formatted as human-readable documents.

electronic image grayscaling An activity outside or in scanning that accurately senses, differentiates, and encodes intermediate shades between black and white in photographs and halftones.

electronic image management (EIM) The techniques associated with recording, storing, retrieving, and transmitting documents by electronic means. See also *digital image, document* and *image.*

electronic image The digital representation of a document.

electronic imaging See *electronic image management.*

electronic mail Software that enables systems to send messages to one another and to be signaled when messages are received.

electronic printing Printing using a laser beam and toner to reproduce computer-generated images on paper.

electronic still-picture imaging Techniques associated with the capture, processing, storing, retrieving, compression, transmission, reproduction, and display of still-picture images using photographic, electronic, or hybrid methods. Excluded are motion picture, television, VCR, and office document images.

emulation The imitation of a function by a system not originally designed to perform that function.

encoded data (1) Signals converted to coded form. (2) Data converted by the use of a code or coded character set in such a manner that reconversion to the original form is possible.

encryption A security measure involving scrambling or transforming digitized data into a coded form by means of an algorithm. Encryption requires a key for decoding.

enhancement A technique for processing an image so that the result is visually clearer than the original image.

EPROM See *erasable programmable read-only memory*.

erasable programmable read-only memory (EPROM) Read-only memory in which data can be erased by ultraviolet light or other means and reprogrammed with appropriate voltage pulses. See also *programmable read-only memory*.

erasable (deprecated) See *rewritable optical disk*.

error detection and correction (EDAC) An operation that includes all phases of identifying and dealing with data errors, including direct-read-after-write and error-correction codes.

error rate The ratio of the amount of erroneously recorded, read, or transmitted information to the total amount sent.

error-correction code (ECC) A method of data recovery that allows for the full recovery of a single physical block of data or 2,048 bytes. ECC is used by the CD-ROM drive and during CD-ROM premastering.

Ethernet A particular implementation of a bus-type local area network that communicates at 10 megabits per second.

extended binary coded decimal interchange code (EBCDIC) An eight-bit computer code that is used to represent uppercase and lowercase characters and special symbols. This system is not a standard, but it is used extensively in IBM and IBM-compatible large mainframe computers.

facsimile (1) An exact copy of a document. (2) A process by which a document is scanned, converted into electrical signals, transmitted, and recorded or displayed as a copy of the original. See also *document* and *original*.

fax See *facsimile*.

feed time The time required to move or feed a document through the transport mechanism of a device.

fiberoptics A technology that uses light as a digital information carrier.

file server In local area networks, the file server allows users to share and thereby conserve the cost of peripherals (such as printers, modems, and scanners) and to likewise share software. The file server is a device on the LAN where the shared software is stored.

firmware Prewritten programs stored in read-only memory (ROM) circuits.

flying aperture scanner A scanner in which the scanning mechanism is part of the light-detection apparatus.

flying spot scanner A scanner employing a spot of light moving over a document, a microimage, or along the contour of a character in order to read it. See also *scanner*.

font A complete family of a given size of type, including capitals, small capitals, and lowercase, together with figures, punctuation marks, ligatures, etc. Italics are considered a separate font.

G (giga) An abbreviation, pronounced "giga" and sometimes written *GB*, denoting one billion units. Sometimes written *GB*. See also *gigabyte*.

gateway A conceptual or logical network station that serves to interconnect two otherwise incompatible networks, network nodes, subnetworks, or devices. Gateways perform a protocol-conversion operation across a wide spectrum of communications functions or layers.

gigabyte (G) A unit of measure, approximately one billion bytes, expressed as 1,000,000,000 bytes, 1G, 10^9 bytes or 2^{30} bytes.

grayscale An array of adjacent neutral-density areas varying by a predetermined rate or step from black to white and used to expose film to determine its sensitometric curve. Synonymous with gray wedge and step tablet.

groove A continuous channel cut or molded into a surface to provide guidance to a probing device. Optical media made with patterns of molded grooves are called *pregrooved*.

Group 1 A CCITT classification of analog facsimile devices that operate at a speed of six minutes using FM modulation. A four-minute speed is also included in this category as a manufacturer's option. See also *International Telegraph and Telephone Consultative Committee*.

Group 2 A CCITT classification of analog facsimile devices that operate at a speed of three minutes using AM modulation. A two-minute speed is also included in this class as a manufacturer's option. See also *International Telegraph and Telephone Consultative Committee*.

Group 3 A CCITT classification of digital facsimile devices that operate at one-minute speeds and use run-length coding of image material to perform redundancy reduction. These machines can also use bandwidth compression to enhance speed. See also *International Telegraph and Telephone Consultative Committee*.

Group 4 A CCITT classification of high-speed (56K bps) facsimile machines that require error-free communications lines and more extensive compression than Group 3. See also *International Telegraph and Telephone Consultative Committee*.

halftone A technique of reproducing continuous-tone illustrations by photographing the image through an etched screen. The finer the screen (measured in lines per inch), the higher the quality of the resulting negative.

handprint character recognition (HCR) An OCR technology designed to turn images of handprinted characters into ASCII code. See also *intelligent character recognition* and *optical character recognition*.

hard error An error that is uncorrectable by a given error-detection and correction (EDAC) system.

hard read error Physical bits of data that cannot be read by the user since they are usually misplaced or absent from the data bitstream.

hardcopy (1) An enlarged reproduction from a microform, usually on paper. (2) A printed copy of machine output in a readable form, for example, output from a computer printer. Synonymous with *printout.*

hardware Equipment, as opposed to *software.*

HCR See *handprint character recognition.*

hertz (Hz) A measure of frequency or bandwidth; the same as cycles per second.

hole A mark burned through or which deforms the information-bearing layer of an optical medium; distinct from a crater in that a hole has no raised "lip" around its perimeter.

host computer A large, shared computer in which an application or database resides or to which a user is connected.

Huffman coding A data compression technique that assigns shorter bit sequences to frequently occurring symbols and longer bit sequences to less frequent symbols. See also *International Telegraph and Telephone Consultative Committee*, *Group 3*, and *Group 4.*

Hz Abbreviation for *hertz.*

I/O See *input/output.*

icon A symbol or pictorial representation of an object or idea. In graphic interfaces, icons represent the different functions and data available to the user; a mouse or other pointing device is usually used to select the desired operation by pointing to one of the icons on the screen.

ICR See *intelligent character recognition.*

image resolution In electronic imaging, the horizontal image resolution indicates how many times a scanner or printing element stops along the width of a page. Vertical image resolution refers to how many times the scanner or printing element stops or marks down the length of a page.

image rotation The ability to rotate images on reader screens or displays in order to allow projection of all images right-side-up and right-reading.

image In electronic imaging, the digital representation of a document.

in-house A reference to something within the organization. This expression is applied to work such as microfilming or scanning done within an organization, as distinct from work done by a service company. Synonymous with *in-plant.*

information area The area of a document that contains intelligence, usually exclusive of the margin.

information system The logical group of subsystems and data required to support the data needs of one or more business processes.

information Data and documents that have been given value through analysis, interpretation, or compilation in a meaningful form.

input device Any device, such as a keyboard, OCR scanner, or mouse that converts data into electronic signals for processing on a computer system.

input (1) The process of entering information into a system, e.g., a computer. (2) Data entered in a system.

input/output (I/O) The general term for the equipment used to communicate with a computer; also, the data involved in such communication.

integrated information system An information system that combines two or more technologies or subsystems, such as CD-ROM and microcomputer.

intelligent character recognition (ICR) An advanced form of OCR technology that may include capabilities such as learning fonts during processing or using context to strengthen probabilities of correct recognition. See also *optical character recognition.*

intelligent scanner A scanner with additional capabilities, such as the ability to recognize characters (OCR) and symbols automatically, or a scanner that can "learn" to recognize commonly used patterns or symbols and other image processing (such as dithering, edge detection, and barcode detection). See also *scanner.*

intelligent terminal (1) A microprocessor-controlled device that can function as a self-contained microcomputer or as a terminal communicating with a remote computer. (2) A terminal device that can do more than send and receive data.

interactive mode See *conversational mode.*

interface (1) The place at which two systems (such as a microcomputer and CD-ROM) meet and interact with each other. (2) Generically, a common interconnection between two components or functions that do not normally interact. The interface can be a piece of hardware equipment, or a common area of computer storage, or some common instructions shared by two or more programs. (3) In the realm of microcomputers, the circuit board that attaches a particular peripheral device to a microcomputer.

International Standardization Organization (ISO) An international standards organization responsible for the creation of the OSI interface model. See *open systems interconnection.*

International Telegraph and Telephone Consultative Committee (CCITT) An international organization that develops international communication standards. See also *Group 1*, *Group 2*, *Group 3*, and *Group 4*.

JBIG See *joint bilevel image group.*

joint bilevel image group (JBIG) An algorithm standard under development by a CCITT/ISO committee for a bitonal compression algorithm that has potential applications in database management systems composed of black and white halftone photographs and text. JBIG's goal is to be lossless. See also *data compression*, *International Telegraph and Telephone Consultative Committee* and *joint photographic experts group.*

joint photographic experts group (JPEG) An algorithm standard under development by a CCITT/ISO committee for a general-purpose compression technique for color and grayscale image applications. JPEG's goal is to be lossless. See also *data compression*, *International Telegraph and Telephone Consultative Committee* and *joint bilevel image group.*

JPEG See *joint photographic experts group.*

jukebox An automated device for housing multiple optical disks and one or more read/write drives.

K (1) An abbreviation, sometimes written *KB*, denoting one thousand bytes. (2) Shorthand for the prefix *kilo* (1,000), as in *kilometers*. See also *kilobyte*.

Kb See *kilobits*.

keyboarding The act of depressing keys on a computer terminal, microfilm equipment, typewriter, etc.

kilobits One thousand data bits. Often expressed as *Kbits* or *Kb*.

kilobyte (K) A unit of measure of approximately one thousand bytes, expressed as 1,000 bytes, 1K, 10^3 bytes or 2^{10} bytes.

LAN See *local area network*.

land and groove A physical feature of optical disks, applied during manufacture, which defines track locations. Some systems record in the grooves and use the lands between the groves as separators. Other systems record on the lands between the groves and use the unaltered groove to determine position.

laser disk See *optical disk*.

laser printer A printing device that uses a laser beam to generate an image, after which it is developed with toner and fused to paper using heat and pressure.

laser A source of light that is monochromatic (of only one wavelength) and highly coherent (with waves in phase both temporally and spatially). An acronym for *light amplification by stimulated emission of radiation*.

LE designation A rating for the life expectancy of recording materials and systems. The number following the LE symbol is a prediction of the minimum life expectancy in years for which information can be retrieved without significant loss when properly stored, for example, LE-100 indicates that information can be retrieved after at least 100 years of storage. See also *life expectancy*.

legible Capable of being read or deciphered.

life expectancy The length of time that information is retrievable in a system. See also *LE designation*.

local area network (LAN) A data communication network of connected devices within a small area, such as a building or group of buildings.

M An abbreviation, pronounced "mega" and sometimes written *MB*, denoting one million units. See also *megabytes*.

magnetic media A variety of magnetically coated materials used by computers in electronic publishing systems for data and program storage.

magnetic tape A continuous flexible recording medium that has a base material impregnated or coated with a magnetic material, on which data can be stored by selective polarization of portions of the surface.

magnetic-ink character recognition (MICR) The machine recognition of digits printed with ink that can be magnetized.

magnetic-ink characters Typefaces specifically designed to be read by magnetic-ink character recognition scanners.

magneto-optic recording The recording of data using optical means to change the polarity of a magnetic field in the recording medium. Data is erasable and/or rewritable. See also *polarity* and *rewritable optical disk*.

magneto-optical Of or pertaining to the interaction of magnetic fields and light. (ISO)

mainframe A class of computer providing large storage capacity, high-speed processing, and complex data handling capabilities normally supporting many concurrent users.

mark A feature on a disk representing recorded information when it is interpreted in the context of the marks surrounding it, which may take the form of a hole or deformation, pit, bubble, reflectivity change, or other change that can be sensed by the optical system.

master (1) In electronic imaging, the first recording, one from which duplicates can be obtained. (2) The act of creating the first recording.

mastering process The creation of the first recording (the master) in the CD-ROM replication process. See also *compact disk-read only memory* and *master*.

MCAV See *modified constant angular velocity*.

MCLV See *modified constant linear velocity*.

media conversion The process of converting information stored on one type of media to another, such as from magnetic tape to optical disk.

media Materials used to store information and data. See also *LE designation* and *life expectancy*.

megabyte (M) A unit of measure of approximately one million bytes, expressed as 1,000,000 bytes, 1M, 10^6 bytes or 2^{20} bytes.

megahertz (MHz) One million cycles per second.

memory The area of a computer system that accepts, holds, and provides access to information and data.

menu In computer software, a pictorial or typographical list of alternative actions, from which the next function or action is selected. See also *icon*.

MH (MHC) See *Modified Huffman Code*.

MHz An abbreviation for *megahertz*.

MICR reader/sorter A device that interprets magnetic-ink characters and translates them to digital information that can be understood by a computer; the device also sorts the original document based on the unique symbols.

MICR See *magnetic-ink character recognition*.

microfiche scanner A device for scanning microfiche. See also *microfiche* and *scanning*.

microfiche Microform in the shape of a rectangular sheet having one or more microimages, usually arranged in a grid pattern, with a heading area across the top. (ISO)

microimage scanner A unit that digitizes microimages. (ISO)

microimage An image too small to be read without magnification. (ISO)

mil A measurement equal to one one-thousandth (0.001) of an inch (0.0254 mm). For example, 5 mils is 0.005 inches.

million instructions per second (MIPS) A measurement of procession speed.

modem A contraction of *modulator-demodulator*; a device that encodes digital data for transmission as analog signals over a particular medium, such as telephone lines, coaxial cables, fiberoptics, or microwaves.

modified constant angular velocity (MCAV) A characteristic of an optical disk that includes several concentric zones. Each zone is read at a constant angular velocity; crossing from one zone to another involves modifying the velocity.

modified constant linear velocity (MCLV) Describes a disk on which tracks are divided into bands. Within a band, the disk spins at a constant angular velocity, but that velocity is different for each band. The relation between velocity and band location is similar to the velocity-radius curve for CLV operation.

Modified Huffman Code (MH or MHC) A run-length code, adopted by the CCITT to perform redundancy reduction, which only removes horizontal redundancy from the image. See also *run-length code*.

modified read code A two-dimensional digital coding scheme for improved transmission speed and quality.

multimedia The combination of graphics, images, sound, animation, and video in a single presentation software program or single presentation system.

National Institute of Standards and Technology microcopy resolution test chart A chart containing a graded series of blocks, made up of lines and spaces, used for determining the optical performance of microfilm equipment and the resolution characteristics of materials used in microrecording.

NBS chart See *National Institute of Standards and Technology microcopy resolution test chart*.

network architecture A philosophy and organizational concept for enabling communications between data processing equipment at multiple locations. The network architecture specifies the processors and terminals, and defines the protocols and software that must be used to accomplish data communications.

network topology The configuration of nodes in a communication network. See also *network*, *node*, *ring network*, *star network*, and *token-ring network*.

network The communications pathway between computers or between terminals and computers. See also *local area network* and *node*.

NIST chart See *National Institute of Standards and Technology microcopy resolution test chart*.

node The termination point for two or more communications links. A node can serve as the control location for forwarding data among the elements of a network or multiple networks, as well as performing other networking and, in some cases, local processing functions. See also *bus network, network, ring network, star network*, and *token-ring network*.

object program The fully compiled or assembled program that is ready to be loaded into a computer. (ISO)

object-oriented program (OOP) A method of programming that views programs as a collection of autonomous agents called *objects*. Each object is responsible for specific tasks.

object-oriented programming system (OOPS) A group of object-oriented programs coupled together. See also *object-oriented program.*

OCR indexing A technique that recognizes lettering, numerals, and other characters and converts them to ASCII (attributes of a document) for subsequent retrieval.

OCR See *optical character recognition.*

ODA See *office document architecture.*

OEM See *original equipment manufacturer.*

off-the-shelf Commercially produced software programs.

office document architecture (ODA) An architecture that can be used to describe a large variety of documents. It allows open interchange of documents in processable (revisable) form, formatted (page image) form, or both. Documents may contain character coded text, raster images, and geometric (vector) images. ODA is being extended to include audio and dynamic images. Extensions will also provide for hypermedia.

offline (1) Pertaining to equipment or devices not under direct control of the central processing unit. (2) Pertaining to equipment or devices not directly linked to a computer-output microfilmer. See also *online.*

online (1) Pertaining to equipment or devices under control of the central processing unit. (2) Pertaining to a user's ability to interact with a computer. (3) Pertaining to equipment or devices under control of a computer-output microfilmer. See also *offline.*

OOP See *object-oriented program.*

OOPS See *object-oriented programming system.*

open systems interconnection (OSI) A seven-layer computer interface model promulgated by the ISO. Each of these layers provides a segment of the overall communications function required for distributed applications.

operating system The collection of programs that, taken together, manage the hardware and software; the operating system makes the hardware usable by providing the mechanisms that application programs use to interact with the computer.

optical character recognition (OCR) A technique by which characters can be machine-identified, then converted into computer processable codes (such as ASCII or EBCDIC).

optical characteristics The properties of an optical system, such as field of view, magnification, brightness of image, image quality, and correction for aberrations.

optical disk A medium that accepts and retains information in the form of marks in a recording layer, which can be read with an optical beam. See also *compact disk-read only memory*, *rewritable optical disk*, and *write-once read-many optical disk.*

optical drive A machine for reading or writing a data storage medium (disk, tape, card, etc.).

optical fiber Thin strands of glass designed for light transmission.

optical memory Memory in which data is recorded and read by optical means. See also *memory*.

optical scanner A scanner that uses light for examining patterns. Optical scanners are often used in pattern recognition or characters recognition. (ISO)

optical scanning A technique to recognize printed or written data and generate a digital representation for storage, transmission, etc. See also *optical character recognition* and *raster scan*.

optical storage A storage system that uses light, specifically light generated by lasers, to record and retrieve information.

optical system The essential and accessory optical elements designed to contribute to the formation of an image.

optical tape Optical memory in tape form.

original equipment manufacturer (OEM) A manufacturer of equipment sold to distributors and others, and marketed under that party's name.

original The document to be reproduced. (ISO)

OSI See *open systems interconnection*.

output device A device, such as a display, printer, or plotter, that converts electronic signals to human-readable form.

output (1) The process of exiting information from a system. (2) Information and data exiting the system.

P An abbreviation, sometimes written *PB*, denoting one quadrillion bytes. See also *petabyte*.

page scanner A scanner that primarily reads and digitizes ordinary business size documents.

parity bit The redundant binary digit added to a series of such digits and assigned the value required to make the sum of all bits odd or even. See also *parity*.

parity check A test that helps ensure the validity of data by determining whether the number of zeros and ones representing a byte of data is odd or even.

parity A method of detection of single-bit errors by adding a bit to the user's data, and setting that bit such that the number of one-bits in all the data is even or odd, normally odd. Sometimes used informally to refer to any error detecting or correcting technique that adds a percentage of redundant information to the user's data.

PC An abbreviation for *personal computer* often used to refer to IBM and IBM-compatible microcomputers.

pel See *pixel*.

peripheral equipment Supplementary equipment external to a computer that puts data into or accepts data from the computer, such as disk drives and printers.

permalloy A metal alloy that is very easy to magnetize and has very low hysteresis. Permalloy is widely used in magnetic reading and writing heads.

permanent record Any records determined by the appropriate public records authority to have sufficient value to warrant their permanent retention and continued preservation.

petabyte (P) A unit of measure of approximately one quadrillion bytes, expressed as 1,000,000,000,000,000 bytes, 1P, 10^{15} bytes or 2^{50} bytes.

phase change recording The ability of a media to transform between amorphous (structureless) and microcrystalline (structured) states. Such media can be transformed a finite number of times.

photomultiplier tube An electronic device that converts light into electrical energy and amplifies it.

pit Broadly used to refer to data-carrying marks in optical media, but originally coined to describe the rimless troughs written in photo-resist on videodisc masters and transferred by molding to videodiscs.

pixel The smallest element of a display surface that can be independently assigned color or intensity.

plotter An output unit that directly produces a hardcopy record of data on a removable medium in the form of a two-dimensional graphic representation. (ISO)

polarity The change or retention of the dark-to-light relationship of an image.

polling A method of controlling the sequence of transmission by terminals on a multipoint line by requiring each terminal to wait until the controlling processor requests it to transmit.

port The connection point on a CPU through which a computer workstation or another device can communicate with the CPU directly.

pre-write defect ratio (1) The ratio of the pre-write defect area in the recording zone to that in the total recording zone. (2) The Ratio of the sum of the time intervals from the beginning to the end of drop-ins on the disk to the total available recording time on the disk.

pre-write defect (1) An area of the recording zone which, when read prior to any recording, yields a signal amplitude greater than (or less than, as expected from the class of media) a specified threshold level. Also referred to as *drop-in*. (2) A defect in the groove or format of a pre-grooved media detected in the media prior to any write on the disk.

premaster An intermediate recording from which a master will be created.

premastering (1) Includes the conversion of digital code, the addition of error correction codes, and the intelligent preprocessing of the data records. (2) The phase of CD-ROM production in which machine-readable and bitstream data are converted to the CD format. See also *compact disk-read only memory* and *premaster*.

printing The process of producing hardcopies.

private line See *dedicated line*.

programmable read-only memory (PROM) Solid state memory for storing programs. Generally, PROM cannot be altered once it has been programmed. See also *erasable programmable read-only memory.*

quality assurance Similar to quality control, but with the program independent of the production organization. See *quality control.*

quality control (1) Planned systematic activities necessary to ensure that media, component, module, or system conforms to established technical requirements. (2) All actions taken to ensure that a development or organization delivers products that meet performance requirements and adhere to standards and procedures. (3) Policy, procedures, and systematic actions established in an enterprise for the purpose of providing and maintaining some degree of confidence in data integrity and accuracy throughout the lifecycle of the data, which includes input, update, manipulation and output. See also *quality assurance.*

RAM See *random access memory.*

random access memory (RAM) Semiconductor memory circuits that hold programs and data during use in a computer system.

raster data An array of pixels that corresponds to the original scanned image.

raster graphics A method of representing a two-dimensional image by dividing it into a rectangular two-dimensional array of picture elements. See also *image* and *pixel.*

raster image An image formed by modulating the intensity of the individual picture elements within a raster array. See also *image*, *pixel*, and *raster.*

raster scan A method of generating or recording the elements of a display image via a line-by-line sweep across the entire display surface, such as the generation of a picture on a television screen.

raster The description of a rectangular or square array formed by a number of horizontal scan lines comprising a number of picture elements. The number of scan lines establishes the vertical dimension of the array, and the number of picture elements forms vertical rows which establish the horizontal dimension of the array.

raster-to-vector conversion Changing of an electronic image from a raster to a vector format. See also *raster*, *rasterization*, and *vector data.*

rasterization (1) The process of converting text or images from an original source document into raster data via scanning. (2) The conversion of information stored as a series of vectors (as found in many engineering drawing systems) or other codes into raster data for display or printing.

read-only memory (ROM) Memory that contains prewritten programs or data. The content of ROM is permanent; the content of random access memory (RAM) is volatile. See also *random access memory.*

reader-scanner A device that scans an image on a microform and produces a bitstream (digital) output, which can subsequently be displayed or printed locally or at a remote location.

realtime An operating mode under which receiving and processing the data and returning the results occurs so quickly that it guides physical processes or interacts instantaneously with a user.

receiver A system component that accepts incoming signals from a transmitting device.

reengineering The process of rethinking and reorganizing the way an information management business process is executed so that additional benefit may be achieved from electronic imaging technology, instead of simply using imaging technology to mimic an existing paper or microform-based operation. Often called *business process reengineering* or *business process redesign*. See also *workflow*.

relational database A database that uses multiple forms and multiple disk files to store information. Records in one file are linked to separate files with related records.

remote access Communications with a computer from locations that are geographically removed from the computer.

reporting Listing information within the database, either on the screen or in printed format.

resolution (1) The relative degree of an image's visual acuity. See also *acuity* and *image*. (2) In electronic imaging, the specific pattern and number of pixels sampled. See also *pixel*. (3) A measure of the capability to delineate picture detail.

resolve To distinguish between adjacent parts or to distinguish the lines within groups of *n* lines that constitute a test pattern.

resolving power A numeric expression of the ability of an optical or photographic system to distinguish or separate two entities spaced closely together. In micrographics, it is the product of the number of the pattern resolved in the NIST test chart multiplied by the reduction, and is expressed in line pairs per millimeter.

retrievability Pragmatic access to information as recorded.

rewritable optical disk An optical disk on which data is recorded. The data in specified areas can subsequently be deleted and other data can be recorded.

ring network A network in which every node has exactly two branches connected to it. (ISO) See also *network*, *network topology*, *node*, and *token-ring network*.

roll film scanner A device for scanning microfilm in roll form. See also *roll microfilm* and *scanner*.

roll microfilm Microfilm that is or can be put on a reel, spool, or core.

ROM See *read-only memory*.

RS-232-C An electrical standard for connecting computer devices for transmission of information one bit at a time (serially), normally implemented through a 25-pin connector in which up to 17 pins are defined. Most often used for transmissions from 100 to 20,000 bits per second.

run-length code A method of redundancy reduction (data compression). See also *data compression*.

SASI (deprecated) Shugart Associates Systems Interface. See *SCSI*.

scale A range of various photographic characteristics such as brightness, exposure, and density.

scan size Designations representing specific horizontal and vertical measurements and corresponding margins. There are two standards for scan size documents, one defined by ANSI for American-size documents and drawings, and the other by ECC for European-size documents and drawings.

scan time The total time to convert text or graphical information to electronic raster form.

scanner A device that converts a document into binary (digital) code by detecting and measuring the intensity of light reflected from paper or transmitted through microfilm. See also *binary code* and *document.*

scanning device A mechanism found in certain microfilm readers that permits shifting the film or the entire optical system so that different portions of the microfilm frame or reel may be viewed.

scanning (1) The conversion of human-readable images from paper or microform into a stream of numeric values, called a bitmapped image. See also *bitmap.* (2) OCR scanning is the conversion of printed or other symbolic information from paper or microform into ASCII code. See also *handprint character recognition, intelligent character recognition*, and *optical character recognition.* (3) The systematic examination of data. (ISO)

SCOOP See *self-coupled optical pickup.*

SCSI See *small computer system interface.*

seek error An error in which an optical disk drive's laser is unable to accurately locate user-requested data as a result of physical or mechanical problems such as vibrations, disk surface irregularities, and poor laser functions.

self-coupled optical pickup (SCOOP) An optical drive design that combines the functions of a laser reading device with a photodetector used to accept tracking and focus error signals.

serial transmission The conveyance of a character of information one bit at a time. See also *bit.*

serial One bit following another over a single pathway.

service bureau See *service company.*

service company An organization that is equipped to provide micrographic and electronic imaging services under contract.

SGML See *standard generalized markup language.*

skewing An image condition resulting from physical distortion inherent to a monitor or from errors in document scanning.

small computer system interface (SCSI) An industry standard for connecting peripheral devices and their controllers to a microprocessor; pronounced "scuzzy." SCSI defines both hardware and software standards for communication between a host computer and a peripheral.

source document (1) The form on which an original transaction was captured. (2) An original record with information.

source-document capture The conversion of documents, usually paper, to digital images.

source-document scanning See *source-document capture.*

standard generalized markup language (SGML) A language for document representation that formalizes markup and frees it of system and processing dependencies. (ISO)

standard (1) A document that establishes engineering and technical limitations and applications for items, materials, processes, methods, designs and engineering practices. (2) A fundamental unit or physical constant, such as an ampere, meter, or absolute zero (Kelvin).

star network A radial or starlike configuration of nodes connected to a central node, in which each node exchanges data directly with the central node. See also *network*, *network topology*, and *node.*

stroke generator A method or device for generating characters using short strokes to draw the alphanumeric character in a manner similar to that used in ordinary handwriting. Some computer-output microfilm devices use this technique to draw the characters on the cathode-ray tube (CRT). See also *vector generator.*

stroke (1) The range of motion of the head position in a disk drive. (2) A straight line or arc used as a segment of a graphic character.

switched line A communications path that is connected to the public telephone network.

switched network A multipoint communications pathway with circuit-switching capabilities, such as the telephone network, telex, and TWX.

synchronizing signal A signal used for the maintenance of predetermined speed relations between the scanning spot and recording spot within each scanning line.

synchronous communications The high-speed transmission of contiguous groups of characters; the stream of monitored and read bits is controlled at a clock rate.

synchronous A mode of data transmission in which data is transmitted in blocks of characters. Sync characters are transmitted prior to actual data transmission to establish timing. Commonly used for batch transmission. Contrast with *asynchronous.*

system An organized collection of hardware, software, supplies, people, maintenance, training, and policies to accomplish a set of specific functions.

T An abbreviation, sometimes written *TB*, denoting one trillion bytes. See also *terabyte.*

tag image file format (TIFF) A standardized header or tag that defines the exact data structure of the associated image.

tape drive A mechanism for controlling the movement of magnetic tape past a reading or writing head.

telecommunications Transmission and reception between terminals, or between terminals and computers, of digitized information over telephone lines.

teleprocessing The processing of data that is received from or sent to remote locations by way of telecommunications channels. See also *telecommunications.*

terabyte (T) A unit of measure of approximately one trillion bytes, expressed as 1,000,000,000,000 bytes, 1T, 10^{12} bytes or 2^{40} bytes.

terminal Any device capable of sending or receiving information over a communications channel.

test chart An image comprised of different (or a number of the same) test targets, designed for the test and evaluation of imaging systems. See also *test element, test frame, test pattern*, and *test target*.

test element An image designed for the test and evaluation of imaging systems, such as a specific line of type that is set in a particular point size, a specific collection of horizontal and vertical lines and spaces, or a line that represents a specific frequency. See also *test chart* and *test pattern*.

test frame A microfilm image of a test chart, test target, test pattern, or test element used to check microfilm quality.

test pattern An image comprised of a collection of test elements, such as resolution test elements and frequency test elements, designed for the test and evaluation of imaging systems. See also *test chart* and *test element*.

test target (1) An image comprised of test patterns and/or test elements designed for the test and evaluation of imaging systems. (2) A high-quality form slide used to evaluate computer-output microfilmer quality. See also *test chart, test element, test frame*, and *test pattern*.

text files Files of human-readable data stored as a series of ASCII characters that can be used by many utility programs other than the specific program for/by which they were created.

threshold A setting, generally used in scanning, that determines whether a particular section of the document is white or black.

thresholding The process by which, in a photodetector, the analog gradation of dark to light is recognized by the scanner's detection mechanism to produce digital signals.

throughput (1) In micrographics, the rate at which documents can be processed through a microfilm camera. (2) The number of film frames, amount of film, etc., output by a computer-output microfilmer or processor. (3) In electronic imaging, the time required in a scanning system to convert hardcopy to digital form.

TIFF See *tag image file format*.

tiling A means of handling large electronic images by storing and managing pieces of imagery (generally rectangular) so that the original image plane is completely covered (or tiled) by these pieces.

time sharing A method of using a computing system that allows a number of users to execute programs concurrently and to interact with the programs during execution.

token-ring network A ring network that allows unidirectional data transmission between data stations by a token-passing procedure over one transmission medium such that the transmitted data returns to the transmitting station. See also *network, network topology, node*, and *ring network*.

track jump The action of moving quickly from one track to another nearby.

track The path followed by the write or read beam.

tracking servo A servo used in an optical drive to keep the reading or writing light spot centered on the information track, despite imperfections in the medium and drive mechanism as well as externally imposed shock and vibration.

transducer A mechanism in a scanner that detects and converts light or lack of light into electrical signals.

transmission speed The rate at which information is passed through communications channels, generally measured in bits per second.

transmission The act of sending information to one or more locations or recipients.

transport mechanism A device used to move film, paper, etc. in a prescribed manner.

transport A moving belt or roller device in a camera or scanner that moves documents through the machine.

turnkey system An integrated configuration of preselected hardware and prewritten software designed to accomplish a particular information processing task.

ultraviolet Pertaining to or designating those radiations that lie beyond (wavelengths shorter than) the blue end of the visible spectrum, approximately from 200 to 400 nanometers.

UNIX A computer operating system developed by Bell Labs with variants supported on a wide range of systems by numerous vendors.

VAR A value-added reseller.

vector data The digital description of an image stored as a series of points and mathematical functions to describe the geometric figure, such as a line, circle, or arc.

vector editor A subsystem (workstation with system software) that is used to manipulate or edit the vector image.

vector generator A method or device for generating graphical information using direction and line length. See also *stroke generator*.

vector-to-raster conversion The changing of an electronic image from vector to a raster format. See also *raster*, *rasterization*, and *vector data*.

vectorization A method of image processing such that alphanumeric characters, lines, drawings, or sketches are converted from raster code to vectors.

video disk A rigid, random-access storage medium for analog or digital information written or read by a laser. See also *optical disk*.

video display terminal (VDT) See *cathode ray tube*.

videotex A system that permits a user to view textual information on an appropriately equipped television set. Videotex systems are subdivided into teletext and view-data systems, whose transmission media consists of broadcast signals and internal computer lines, respectively.

WAN See *wide area network*.

warping angle The maximum angle between the axis of rotation of the disk and the normal to the reference plane when the disk is properly clamped to the spindle. The measurement should be made with the disk at a specified rotational speed.

wave A pattern of propagation of certain kinds of energy, including light. Waves have several properties, notably amplitude, frequency, wavelength, phase, and, in some cases, polarization. The frequency of a wave is the number of times it cycles through one wavelength in a given period of time. Frequency is expressed in hertz or cycles per second.

wavelength The length of a wave measured from any point on one wave to the corresponding point on the next wave, usually from crest to crest. Wavelength determines whether radiant energy is classed as gamma rays, X-rays, ultraviolet, visible, infrared radiant energy, or radio waves. The wavelength of visible radiant energy is the chief determinant of its perceived color.

wide area network (WAN) A communications network that links broad geographic areas. See also *local area network* and *network*.

workflow In imaging software, a program that queues, tracks, and otherwise manages documents and collections of documents as they progress from entry into the system, through the various departments in the organization, to their final destination.

WORM See *write-once, read-many*.

write To record information on a storage device, a data medium, or an output display.

write-once, read-many (WORM) A digital optical disk on which data is recorded by the user once and can be read many times.

zoom To enlarge the presentation size of information (such as text or images) displayed on a workstation monitor.

Index

I